Algal Flora of Paddy Fields of Vidarbha Region, Maharashtra

Algal Flora of Paddy Fields of Vidarbha Region, Maharashtra

Dr. K.J. CHERIAN

BIOTECH
2012
BIOTECH BOOKS

ISBN 978-81-7622-255-6

Published by : **BIOTECH BOOKS**
4762-63/23, Ansari Road,
Darya Ganj,
New Delhi - 110 002
Phone: +91-011-23262132
E-mail: biotechbooks@yahoo.co.in

Laser Typesetting : **Sushil Kumar**

Printed at : **Chawla Offset Printers**
Delhi - 110 052

PRINTED IN INDIA

PREFACE

Studies on soil algae gaining its importance due to its contribution to the conditioning of soil for an effective growth of plants. The use of various inorganic chemicals for the protection & nutrition especially to the crop plants, the biodiversity of the microorganisms of the crop fields are in danger. It is necessary to study & enlist the type of microorganisms of the soil to know the biodiversity of each part of India.

The present work enlighten the readers about our biodiversity strength of microorganisms belongs to the algal groups. It also helps the young researchers with the methodology of collection, culturing & identification of algal components of soil. Bioevaluation of the soil is a must before going to any type of innovative approach in the cultivation. Isolation & cultural studies in soil microorganisms helps us to know the role of each microorganisms in the soil. Generally these soil algae prevent soil erosion, increase humus & soil mineral ions. This volume establishes the algal components of paddy field soil in Vidarbha region & also its biodiversity strength.

I hope this work will enlighten the biologists especially the young researchers for further studies in this field & makes a concern for biodiversity protection.

I heartly thank to Prof. D.R. Khanna, Head, Department of Zoology and Environmental Science, Gurukula Kangri Vishwavidyalaya, Haridwar, India for his encouragement and support for the completion of this book. I also thank Biotech Books & Publishers, Delhi in bringing out this volume in time with nice presentation.

CONTENTS

Chapter 1

INTRODUCTION

Paddy is cultivated in an area over 80 million acres in India and it forms the major staple food of the people. It is therefore natural that our efforts are directed towards increasing its production by introducing high yielding varieties, generous manurial practices, control measures for pests and diseases and other modern agriculture practices. Yield of the paddy crop depends on the soil.

Soil is a very complex system, composed of two parts as living and nonliving parts. The non-living part composed of inorganic material derived from rocks and organic matter derived from plant residues, water, air etc. The living part has microfauna and microflora. The microflora comprises of algae, fungi bacteria and viruses. The microfauna mostly consist of protozoa, nematodes and arthropods. The living constituent of the soil continue their activity year after year. The role played by the micro organisms is of paramount importance as they bring about chemical and physical changes in the soil nature by various reactions. Some of these reactions such as decomposition of organic remains and release of carbon and nitrogen are significant in the soil eco-system. Inorganic compounds of iron, phosphorus, manganese etc. which are not readily absorbed by the higher plants are made available to them by the oxidation-reduction reactions of the soil micro-organisms. The product of oxidation and reduction reactions, apart from exerting a solvent effect on title parent material of the soil (Martin, 1949; Chodat, 1928), affect the inorganic fertilizers added to the soil too. The combined effect of all these reactions is that the soil becomes a fertile medium for the growth of plants. The study of soil micro-organisms by various workers envisages the role of micro-organisms in relation to the soil conditions and ecology of the soil (Fritsch, 1907).

The presence of algae over soil were reported from all the continents by different workers. Naegeli (1849), Kuetzino (1888), and Graebner (1895) were the earliest workers who reported about the occurrence of algae over soil. In 1888 Treub reported the significance of soil algae as the first colonizers of unfavourable substrate. The important workers in the first part of the twentieth century were Fritsch (1907 a,b; 1922 a,b; 1936), Bristol (1919, 1920, 1926, 1927a,b; 1928) Gistl (1932, 1933), Peterson (1935), Fritsch and De (1938), Fritsch and John (1942) and Lund (1947). The first soil algae isolated into an auxenic culture originated from a soil collection by Beijernick (1893). Thus algae have been recognised as a component of the soil microbial community since the late nineteenth century. Only during the last decade have algae received recognition as an integral part of the soil ecosystem. The importance of algal growth in the colonization of new ground, formation of humus were discussed by Fritsch (1907). Algae are equally important as pioneers which for the initial community in plant succession on certain eroded soils (Booth, 1941). Basic concepts of the morphology, distribution and taxonomy of algae were developed through studies of terrestrial algae by Fritsch (1922), Fritsh and Hanes (1923), Bristol-Roach (1927), Peterson (1933, 1934), James (1935), John (1942) and Luna (1946a,b, 1947). They laid the foundation for later studies on Physiology and ecology of soil algae.

Bristol (1919) studied the viability of algal spores in stored soils. She reported the growth of a number of Cyanophyceae and Chlorophyceae in the cultures of soils stored for 23 to 73 years. Later (1920), she observed that the luxuriance of algal flora in the soils was influenced by available moisture and nutrients. Fritsch (1922) made studies on the moisture relations of terrestrial algae to determine their capacities to withstand long periods of desiccation. Bristol (1926, 1927a,b) found that soil algae vary seasonally and that pH and carbon compounds had a marked effect on the growth and distribution of algae in some normal English soils. In 1928 she published a paper on the distribution of algae in the horizontal stratifications of the soil. Martin et al. (1955) and Lund (1947) were also worked out the chemical, physical and physiological changes brought about by various microbial activities, which effect on the structure and fertility of soil.

Gistl (1932) studied the effect of manuring on algal populations of the soil and showed that physico-chemical conditions of a given algal flora and the appearance of individual species. In 1933 he showed that application of fertilizers for a long time, increased the number of algal species growing on the soils. Petersen (1935) investigated the algal flora of soil samples in Denmark, from various formations such as heath, meadow and manured pasture. He found that cultivated soils contained maximum algal species. Later in 1935, Petersen reviewed the work on soil algae and discussed the factors determining their occurrence in the soil. He concluded that the algal species were carried down to different depths in the soil due to percolation. He considered epiterranean algae more important than subterranean algae.

Fritsch and John (1942) found a correlation between the composition of the algal flora and the soil characteristics. Blue-green algae showed a distinct preference for soils with alkaline pH. Lund (1947) investigated the flora of 54 diverse soils by means of cultures and direct observation and concluded that blue green algae predominated in the non-base deficient soils, and green algae favoured base deficient soils. He also observed that the number of algae in these soils varied with the weather conditions. Fletcher and Martin (1948) tried to study the effects of algae in the rain crusts of desert soils of Arizona. They found that soil algae contributed appreciably to the nitrogen status of those soils. The amino or other nitrogeneous compounds formed by soil algae are released into the surrounding medium, and thus the algae constituted a renewable supply of soil nitrogen.

The development of soil algae in sodium padzolic soils and their relation to the cultivated plants were studied by Shtina (1957). She observed leguminous plants favoured the development of green algae and the cereal crops supported the growth of blue-green algae and diatoms. Durrell and Shields worked out the algal flora of the large number of soil sample from U.S.A. particularly from Colorado, Nevada test sites and some tropical soils in South America. Durrell (1959) studied 223 Colorado soils, representing irrigated and dry farm land, grass land, mountainous and forest areas. He obtained 85 algal species belonging to 40 genera from these soils. *Phormidium, Oscillatoria* and *Nostoc* common blue-green genera in these soils. Bunt (1961) investigated the algae of the Australian paddy soils. He reported that the paddy soils in North Queensland contained only a small portion of nitrogen fixing algae as compared to the total algal flora. Durrell and Shields in 1962, reported the algae occurring in the Death Valley, California, with special reference to those found in the soil. He studied 113 samples, taken from several transacts, representing various geological and ecological conditions. Sixty one samples from his collection contained algae. In the following year (1963), he studied 120 samples from some tropical places like Gulf of Mexico, Ecuador and Columbia took an inventory of the algae occurring in them. He found 16 species belonging to green and 46 belonging to blue-green algae. Shields, in a joint publication with Drouet (1962), studied the distribution of algae within the Nevada test site. They were of the view that soil texture, which influences cohesiveness and moisture levels, strongly affects the composition of terrestrial algal populations.

Lund (1962) and Shields and Durrell (1964) reviewed the taxonomic as well as ecological investigations concerning soil algae, carried out in different parts of the world by various workers. The possibility of using blue-green algae as green manure was studied by Watanabe (1965) in Japan. Jurgenson and Davey (1968) studied the algal populations from soils representing several major forest types, and the nitrogen fixing blue-green algae of a forest nursery. They reported a total absence of nitrogen fixing blue-green algae in acid forest soils. However, they observed a pronounce development of these algae on the surface of acid nursery soils. Similar results were obtained by them from the Alakan tundra soils. Granhall and Henriksson (1968) reported that only 47 percent of Swedish soils harbour nitrogen fixing blue-green algae. Determining factors for the distribution of soil algae were soil pH, humidity in the soil, light and temperature. Blue-green algae, like Anabaena and Nostoc, were reported to play and important role in the soils in polar and subpolar regions. Shtina (1969) published the results of her experimental studies in algology. She discussed in her paper, the regular patterns of algal Clistribution ane their role in soil forming process. In her studies, she used the species of blue-green algae which occurred in particular soils as soil indicators. Granhall (1970) reported that *Nostoc* and *Anabaena* were the commonest nitrogen fixing algae in Swedish soils and had the greatest pH tolerance. Trainor (1970) studied the survival of algae in a desiccated soil from a Connecticut cornfield. He found 11 species belonging to 9 genera. In 1971, Khoja and Whitton published the results of survey, on the ability of 24 strains of filamentous blue-green algae to grow in the dark. Shtina (1972) reported some peculiarities in the distribution of nitrogen fixing blue-green algae in Russian soils. She observed that during cultivation, the algal communities change; in particular, the proportion of nitrogen fixing blue-green algae increases. Presence of calcium and slightly acidic or neutral soil pH promote the growth of nitrogen fixing algae, while high doses of mineral nitrogen suppressed their growth. The growth of nitrogen fixators wac also influenced by root excretion of the crop plants.

Roger and Jacq (1972) and Roger (1973) studied the distribution of blue-green algae in several paddy soils of Senegal and estimated nitrogen fixation in a variety of soil types, including sandy paddy soils, sugarcane fields and other neutral and acidic soils. Brock (1973) made observations on a wide variety of acidic environments and found that blue-green algae were completely absent from habitats in which the pH was less than 4-5. Dooley and Houghton (1973a,b) too stated that the occurrence of blue-green algae in peat soils was can trolled by the pH level and the presence of phosphates in the environment. The micro population of the soil samples studied tended to increase in size and diversity with increasing plant growth and availability of moisture.

El-Nawawy and Hamdi (1975) reviewed the work done on soil blue-green algae in Egypt. They observed that *Nostoc*, Anabaena and *Calothrix* were dominant among a wide variety of blue-green algae in Egyptian soils. Alexander (1975) reviewed the work done on nitrogen fixation by blue-green algae in polar and sub-polar regions. Renaut et ale (1975) investigated over 500 soil samples from various terrestrial habitats in Morocco for nitrogen fixing blue-green algae. The common genera in the soils of Morocco were *Oscillatoria, Spirulina, Lyngbya, Microcolous* and *Nostoc*. Roger and Reynaud (1976) studied the algal population dynamics in a Senegalese rice field during growing season. De Halperin (1976) and De Halperin et ale (1976) studied the algal crusts of sub-humid and semi-arid soils of Chaco and Formosa provinces in Argentina. Blue-green algae were found dominating in the algal crusts and she described 48 taxa belonging to 11 genera. Englund (1976) reported that *Anabaena* and *Nostoc* species fixed appreciable amounts of nitrogen on Heimaey Volcano, 18 months after erruption. Kaszubaik (1976) studied the influence of mineral fertilizers on some groups microorganisms at different soil moistures. Tell (1976) investigated the soil algae around Buenos aires. Jones (1977a,b,c,d) studied nitrogen fixation by blue-green algae in the subtropical grass lands. King and Ward (1977) studied soil algae of disturbed and undisturbed soils. They concluded that disturbed of soil and soil pH were the primary factors which influenced algal distribution.

The Propagation of blue green algae will not only enrich the nitrogen status of the soil by their nitrogen fixation process, but also provide organic matter and other biologically potent Substances for plant growth (Fogg, 1962; Singh, 1961), Lund

(1962) has reviewed the literature on soil algae with reference to their physiology and biochemistry whereas Shields and Durrell (1964) reviewed them in relation to soil fertility. Blue-green algae have been shown to excrete organic nitrogenous substance amounting to as much as 50% of the total nitrogen fixed and some of them are in the form of polypeptides (Fogg,, 1951, 1960), as free amino acids (Watanobe, 1951; Venkataraman and Saxena, 1963; Stewart, 1963; Venkataraman et al., 1964). Lehri and Mehrotra (1970), Watanabe (1970) and Venkatraman (1972) were concluded that nitrogen fixing blue green algae make the major contribution to the soil fertility. Besides releasing oxygen for paddy roots (Harrison et al., 1913) and increasing soil phosphate (Fuller and Rogens, 1952), some of these algae seems to prevent the loss of soil ammonia bleaching out of nitrates by conversion them into organic nitrogen (Waksman, 1952). The efforts made to increase the soil fertility by utilizing microbes have been amply rewarded (Subrahmanyam, et al., 1965; Venkataraman and Goyal, 1968, Venkataraman, 1975).

Till the dawn of 20th Century, bacteria were thought to be responsible for most of the soil process, and other constituent groups witch as fungi and algae were not given due consideration in the systematic study of the soil populations. During the last few decades, it has been established that besides bacteria, fungi and algae also have a remarkable effect on the fertility of the soil. In the light of these facts, soil fungi have been worked out extensively but algae have remained relatively neglected. It is only recently that the ecological importance of algae has been gradually recognized, and it is now accepted that their taxonomic, ecological and physiological study is necessary for greater food yield from the cultivated fields all over the world.

Soil algae exert profound beneficial effect on physical and chemical properties of the soil supporting the crop. They increase the organic content of the soil upon death and decomposition (Fletcher and Martin, 1948). Blue-green algae makes a major contribution to the soil fertility by atmospheric nitrogen fixation (Venkataraman, 1961; Fogg and Stewart, 1965; Stewart, 1970, 1973).

Algae also produce surface humus after death (Aleksandrova, 1951) and exert solvent action on certain minerals (Chodat, 1928; Martin, 1940), maintaining a reserve supply of elements in a semi available from for higher plants (Allison et al., 1973). The extra cellular products secreted by these algae also help in the growth of the plants and increase the crop yield (Venkataraman, 1961; Watanabe, 1961; Singh, 1967). Soil algae checks the soil erosion by their soil aggregating capacity (Marathe, 1972a). They conserve water in soil (Marathe, 1970).

In early twentieth century itself work on soil algae were started. In the beginning of the twentieth century Lyengar (1929) reviewed the work carried out on soil algae. Soil algae of soil samples of Lahore was studied by Singh in 1933. He observed the predominance of blue-green algae and most common genera were found to be Oscillatoria, Phormidium, Nostoc and Anabaena. Banerrji (1935) investigated the algal flora in culture of desiccated and muddy soils from rice fields of Fariopur, East Bengal. He obtained a total of 9 algae, out which 8 were Myxophyceae and 1 chlorophyceae. Singh, R.N. (1939), while studying the algae in cultures of the rice fields of four districts of Uttar Pradesh, viz. Mirzapur, Varanasi, Gorakhpur and Basti, recorded 17 species of Chlorophyceae, 14 species of Myxophyceae and 12 species of diatoms. De (1939) isolated phormidium foveolarum and 4 new species of Anabaena from the paddy field soils of Faridapur, East Bengal.

Singh (1942) studied soil algae form ecological point of view. He found that algae show a marked degree of adaptation to the conditions of desiccation, to which they are exposed. He came to the conclusion that an alkaline medium, high temperature, combined with high atmospheric humidities are responsible for the abundance of blue-green algae in Indian soils. In 1949, Gonzalyes and Gangla studied the algal flora of paddy field soils from 5 places around Bombay and observed a luxuriant growth of algae in the cultures. There was a predominance of Cyanophycea in the soils studied by them. The organic content and the water holding capacity of the soil could be considerably increased by reclamation of usar lands by soil algae (Singh, 1950). Work on paddy field soil algae of Allahabad (Mitra, 1951) revealed the dominance of Cyanophyceae over other groups. Dutta and Venkataraman (1958) recorded 17 blue-green algae and 3 green algae from cultivated and noncultivated soils. The rhizosphere algae of different crop plants were studied by Gonzalves and Yalavigi (1960).

A monograph on soil algae by Singh (1961) covered different aspects of soil algae and its capacity. The effects of fertilizers on soil algae of paddy fields from Karjat, near Bombay were studied by Harathe (1963) and Marathe and Navalkar (1963). They found that phosphate exerted a maximum effects on soil algae. Marathe (1964) also observed that green manure stimulated the growth of a number of algae, especially *Nostoc muscorum*. The inoculation of algae viz., *Nostoc, Anabaena, Microcustis, Phormidium,* Aphanothece etc. increased the fertility of soil (Relwani, 1963).

The algae of paddy fields from soils of Ballia and Ghaziapur districts of Uttar Pradesh were studied by Pandey (1965). He described a total of 107 algae, out of which 74 were blue-green. Higher percentage of Cyanophyceae was observed in mangrove swamps near Bombay by Marathe (1965). Application of lime stimulated the growth of blue-green algae, especially *Nostoc* and *Anabaena* (Amma et al., 1967). Subramanian et al. (1965) surveyed the experimental rice plots of I.A.R.I., Cuttack, and reported 17 species of blue-green algae. In a survey of the nitrogen fixing algae in the rice soils of Madras state sponsored by I.C.A.R. (1965), 169 species belonging to 31 genera of Cyanophyceae were reported. *Nostoc, Cylindrospermum, Anabaena* and *Aulosira* were the dam ment forms in these soils. Gupta (1966) reported 48 species of Myxophyceae and 9 species of Chlorophyceae from paddy fields of Allahabad district. The blue-green algae formed the dominant flora among the soil algae in certain soils of Andhra Pradesh (Subba Raju, 1967).

The variations in pH and moisture contents of soils during different seasons are the determining factors for the luxuriance of algal growth (Prasad and Srivastava, 1968). The nitrate deficient rice fields supported a rich growth of nitrogen fixing blue-green algae (Gupta and Kumar, 1968). The blue-greens dominated other groups in Paddy field soils studied by Khurana and Venkatararnan (1968). Laloraya (1968) too studied extensively the Blue-green algal flora of the paddy field soils in India. In the subsequent years, Marathe (1967, 1969, 1970, 1972a and b), Marathe and Anantani (1972) and Anantani and Marathe (1972a and b) had contributed to the taxonomical and ecological study of the cultivated fields, arid and semi-arid soils, etc. Mahapatra et al. (1971) studied the effect of nitrogen phosphorus and potassium fertilizers on the prevalence of algae under field conditions. They observed that phosphorus had the most promoting effect on the growth of algae.

The work done on blue-green algae in relation to agriculture in India was reviewed by Sankaram in 1971. The blue-greens occurring both in cultivated and uncultivated moist soils was reviewed by Subba Raju in 1972. The blue-green algal flora of Jagdalpur was studied by Subramanyam in 1972. Tiwari (1972) reported the blue-green algal flora of the rice fields of India. The algal flora of the hills was quite similar to that obtained from the soils of the plains (Das et al., 1974). From the studies of blue-green algae of paddy fields soil of North-East India and south India by Laloraya and Mitra (1974 a, b and c) concluded that algal flora of a place depended on the manure content of the soil. Singh (1974) studied the Myxophycean algal flora of the paddy fields of sub-mountainous region of the Himalayas. In 1975, Bharati and Bhongale studied the soil algae of Dharwar in Karnataka State. Marathe and Khusaldas (1975) observed that algal crusts had increased the moisture and humus content of the soil and also exerted some effect on pH and phosphorus content of the soil.

Venkatraman (1975) made and All India survey of the soil samples from rice fields. He collected 2213 soil samples from 15 states and studied their algal flora. He reported that only 33 percent of these samples, harboured nitrogen fixing algae. He observed regional differences in the distribution of dominant forms. Khan and Mathur (1976) recorded 24 species belonging to Cyanophyceae and 24 to Chlorophyceae from the rice fields around Dehradun. Rajarammohan Rao (1976) observed that nitrogen fixation in paddy fields under unfolded condition was negligible and that soil submergence accelerated nitrogen fixation. Ali and Sandhu (1976) surveyed saline soils of Punjab (Pakistan) and were able to identify 100 species of blue-green algae, belonging to 27 genera.

Mathur and Gaur (1977) observed that the efficiency of algal inoculation was enhanced in the presence of humus and recorded 4 percent increase in yield over algae alone. Prasad and

Mohrotra (1978) described 16 blue green algae from different crop fields of Uttar Pradesh. Prasad et al. (1978) studied soil samples from 16 types of crop fields for the pH tolerance of blue-green algae. They reported 216 taxa occurring in soils, with a pH range from 5.5-8.5.

The soil algal flora of paddy fields have been reported by different workers form different places. Singh (1939) studied soil algae from 4 samples of U.P., De (1939) from rice fields of Faridapur (now in Bangladesh), Gonsalves and Gangla (1949) made a valuable study of the rice field from Bombay in culture. Mitra (1951) studied the paddy algal flora of certain Indian soils. Khan (1957) from rice fields of Kashmir, Dharmajirao (1968) from Behrampur of Orissa, Singha and Pandey (1972), Kamath and Patel (1973) and Bhongale (1979) from Dharwar have also studied the paddy soil algal flora.

A search through the literature on soil algae, shows its importance to the crops by its effect on soil fertility. Soil algae of paddy field in different parts of India were studied with reference to their taxonomy and ecology. The soil algae of orchard soils of some places in Vidarbha was studied by Choudhary (1977) and blue-green of cultivated soils of Vidarbha was done by Reddy (1979). The present study deals with the algae of paddy fields in Vidarbha region of Maharashtra.

The taxonomy and ecology of soil algae of crop fields were done in different parts of India. In Vidarbha algae of orchad soils (Choudhary 1977) and blue greens of cultivated lands of Vidarbha (Reddy 1979) were worked out by soil culture studies. Studies on the algal flora of a particular field or region may not be complete, just by cultures of soil samples collected only in a particular season. Algal flora of paddy fields in Vidarbha was not yet worked out. The present investigation was therefore undertaken for further studies on different aspect relating to paddy and paddy field alga of Vidarbha region.

The Vidarbha is located in Maharashtra state and one of the 3 divisions of Maharashtra State. It is situated in the Central India and lies between 16°4' and 22°1' north latitude and 72°6' and 80°9' east longitude. It is an agricultural region with most of its area under cultivation of different types of crops. It comprises eight districts (presently 10 i.e Gondia and Gadchiroli) namely, Nagpur, Bhandara (includes Gondia), Chandrapur (includes Gadchiroli), Wardha, Amravati, Yavatmal, Buldana and Akola. The Map-III shows the location of each district in Vidarbha.

Paddy field in Vidarbha shows different habitat in different seasons and at different stages of crop growth. In Vidarbha most of the paddy fields are under water from August to October (Rainy season). It becomes marshy from October to December. From December to February the soil become further dried and forms a moist soil. During summer i.e. from February to June, the soils become fully dried and vegetative stages of algae are usually not observed. This variation of habitat may further vary at different location of the field. Thus for the study of paddy field algae in details and to avoid the possibility of missing any alga from observation, following care was taken.

(a) Samples for soil culture study were collected at different seasons.

(b) Random sampling of the soil at different locations from all paddy fields under study was done and its culture in nutritive media were prepared in replications.

(c) A variety of culture media were used for the same soil sample, to reduce the risk of a species deletion due to the culture conditions.

(d) Fresh algal collections were done directly of those algae which grows luxuriantly in the fields.

Physical and chemical analysis of the soil samples were carried out to show the affinity of a particular alga to that of a particular soil composition. Since pH being one of the primary factors for the algal distribution (King & Ward 1977), the pH of the field area were noted during collection to establish relationship between the pH and distribution of algae.

MAP – I

TOPOGRAPHY OF INDIA SHOWING LOCATION OF MAHARASHTRA

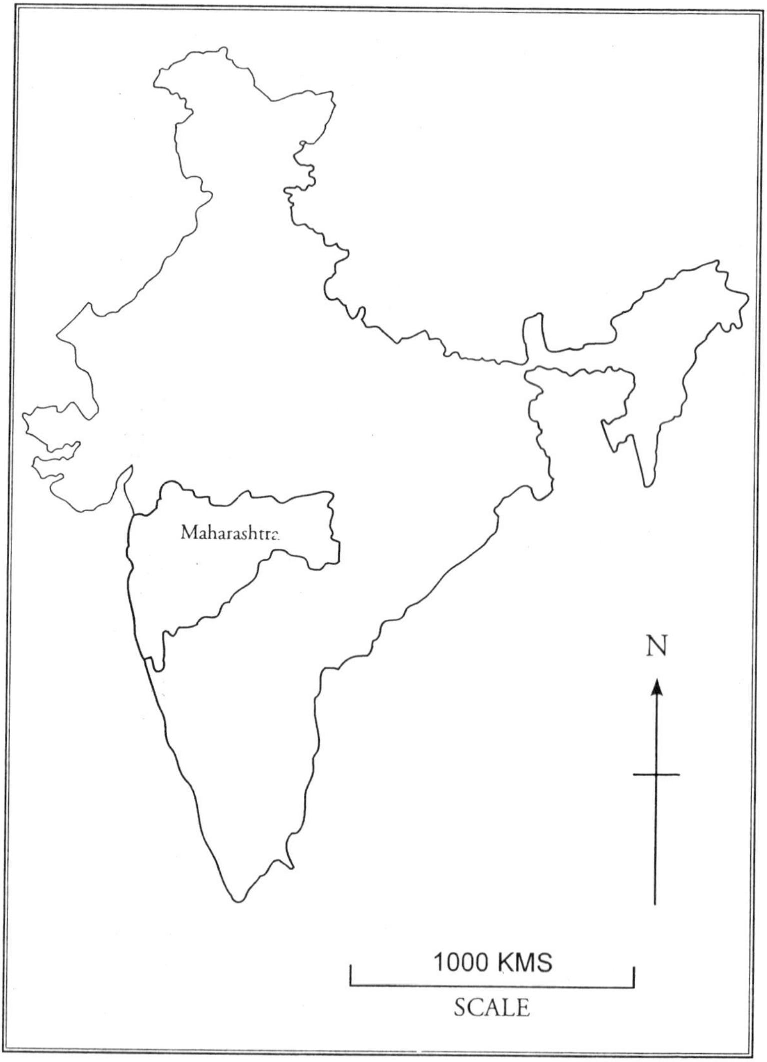

MAP – II

TOPOGRAPHY OF MAHARASHTRA SHOWING LOCATION OF VIDARBHA

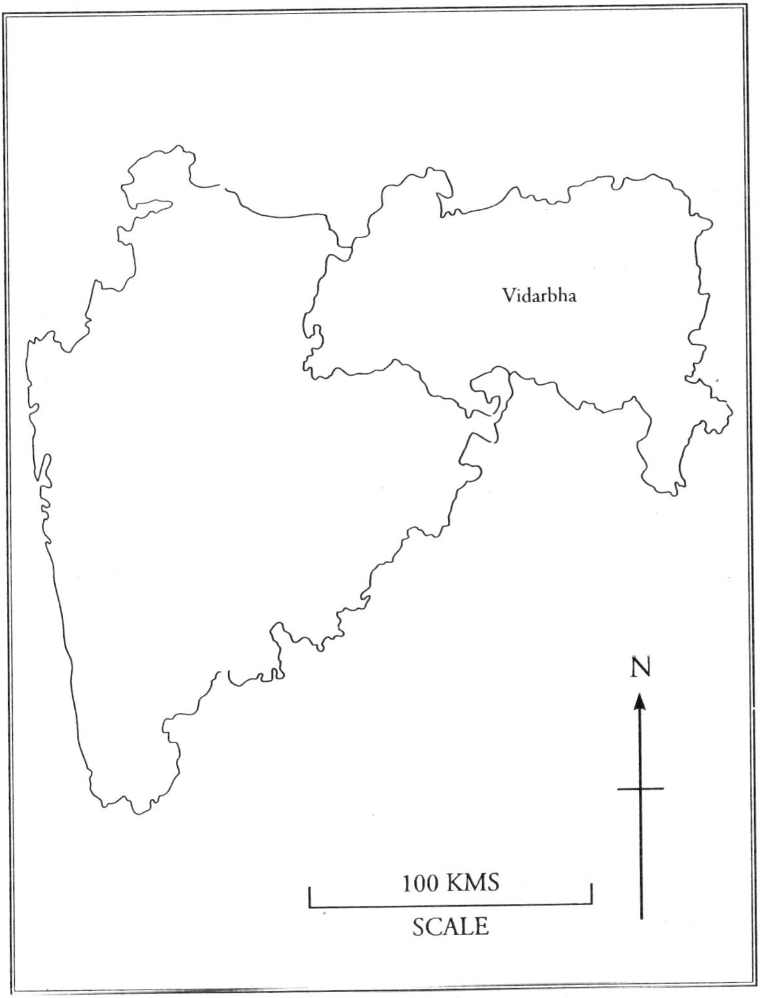

MAP – III

TOPOGRAPHY OF VIDARBHA REGION SHOWING LOCATION OF DISTRICTS

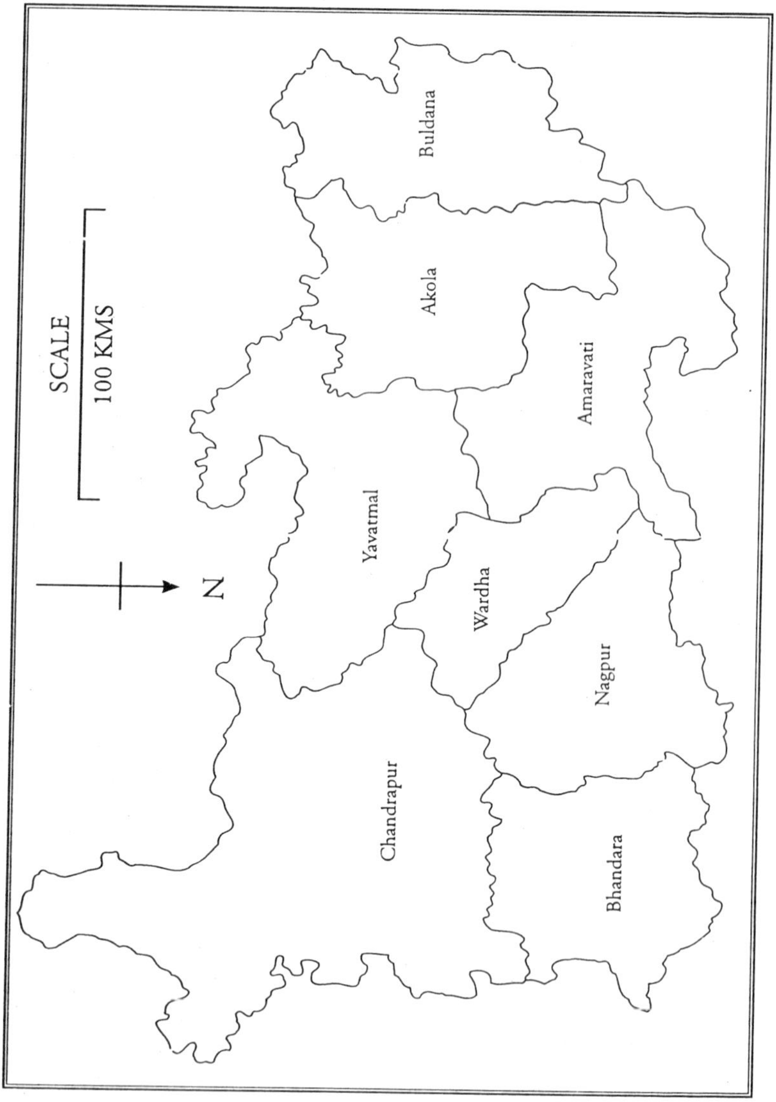

Chapter 2

MATERIALS AND METHODS

A. Collection of soils :

Soil samples were collected in new unused polythene bags from the paddy fields of all the eight districts of Vidarbha. Soil collections were made in summer, before the starting of the crop and after the harvesting of the crop.

In all soil from 53 places belonging to the eight districts of Vidarbha were collected (Table-I). From each place, 5 fields were chosen for the collection of soil samples. A composite soil sample was made by mixing the samples from 5 pits of the same paddy field. Each pit is of a volume of 10 cubic cm. The soil samples were collected with the help of a small rectangular shovel, and the stones and litter were scrapped off with the scalpel.

The shovel, scalpel etc used in the collections were sterilized after each collection and kept completely covered with polythene bags. The soil samples were brought to the laboratory and were dried at room temperature with clean filter paper and kept in labeled packets.

B. Collection of algae from fields :

Frequent visits to paddy fields during rainy and winter season helped me to get maximum algal collections. These fresh collections were made with the help of scalpel, forceps and sieves and were collected in polythene bags. These were brought to the laboratory and identified and developed live cultures of it.

C. Algal cultures :

1. **Culture vessels:** Conical flasks, flat bottles, petridishes were used in the cultures. These glasswares were well washed first with vim powder, then twice with tap water. These were then rinsed with concentrated sulphuric acid and finally washed with distilled water, 3 to 4 times. Then the flasks and bottles were closed with plugs of non-absorbant cotton, whereas the petridishes were closed with its pair petridishes.

2. **Culture Media :** Two types of cultures, liquid and moist, were prepared for studying the soil algae and fresh collections. The bottles and conical flasks were used for the liquid cultures whereas petridishes were used for the moist cultures.

The culturing vessels with culture media were sterilized in an autoclave at 2 lbs pressure for 20 minutes prior to inoculation.

Nine different culture media were employed for culturing. They are as follows:

1. **De's (1939) modified Beneck's medium :**

KNO_3	-	0.2	gm
$Mgso_4\ 7H_2O$	-	0.2	gm
K_2HPO_4	-	0.2	gm
$CaCl_2\ 2H_2O$	-	0.1	gm

$FeCl_3$ (1%)	-	2 drops
EOTA	-	Traces
Distilled water	-	1000 ml

2. Allen and Arnon's (modified) medium (Allen &Arnon, 1955, b):

$MgSO_4\ 7H_2O$	-	0.001 M
$CaCl_2$	-	0.0005M
NaCl	-	0.004 M
K_2HPO_4	-	0.002 M

Minor elements:

$MnSO_4\ 4H_2O$	-	0.5 ppm
MoO_3	-	0.50 ppm
$ZnSO_{4.}\ 4H_2O$	-	0.05 ppm
$CuSO_4\ .5H_2O$	-	0.02 ppm
H_3Bo_3	-	0.50 ppm
$NH_4\ VO_3$	-	0.01 ppm
$CO\ (NO_3)_2{\cdot}6H_2O$	-	0.01 ppm
$N\ iSO_4.6\ H_2O$	-	0.01 ppm
$Cr_2\ (SO_4)_3\ K_2SO_4{\cdot}24H_2O$	-	0.01 ppm
$Na_2WO_4\ 2H_2O$	-	0.01 ppm
$TiO\ (C_2\ O_4)\ x.YH_2O$	-	0.01 ppm
Fe EDTA	-	4 ppm

3. Chu. No. 10 (Modified Medium (Gerloff et al 1950):

$Ca\ (NO_3)_2$	-	0.04 gm
$K_2\ HP0_2$	-	0.01 gm

$MgSO_2 . 7H_2O$	-	0.025	gm
$Na_2 C0_3$	-	0.020	gm
$Na_2 SiO_2$	-	0.025	gm
Ferric Citrate	-	0.003	gm
Citric acid	-	0.03	gm
A_5 solution	-	1.0	ml
Distilled water	-	1000	ml

A5 trace element stock solution :

H_3B0_3	-	2.86	gm
$MnCI_2 4H_2O$	-	1.81	gm
$ZnSO_4 .7H_2O$	-	0.222	gm
MoO3 (85%)	-	0.0177	gm
$CuSO_4 . 5H_2O$	-	0.07	gm
Distilled water	-	1000	ml

4. **Chu. No. 9 medium :**

$Ca(NO_3)_2$	-	0. 02 - 0.04	gm
$K_2 HPO_4$	-	0.002 - 0.04	gm
$MgSO_4 \cdot 7H_2O$	-	0.01 - 0.04	gm
K_2SiO_3	-	0.025	gm
$CaCO_3$	-	0.01	gm
$FeCI_3$	-	0.01	gm

5. **ASM – 1 Medium (Gorham et al 1964) :**

$NaNO_3$	-	20000	micromoles/litre
$Mg CI_2$	-	200	"
$MgSO_4 \cdot 7H_2O$	-	200	"

$CaCI_2 . 2H_2O$	-	200	"
$K_2 HPO_4$	-	100	"
Na_2HPO_4	-	100	"
$FeCI_2$	-	4	"
Na_2 EDTA	-	20	"
H_3BO_3	-	40	"
Mn CI_2	-	41 + 20-7	"
Zn CI_2	-	3.2 "	"
Cu CI_2	-	0.008 "	"
Co CI_2	-	0.08 "	"

6. Brandwein's Solution (Brandwein,1940, 1941)

Difco-Bacto tryptophane	-	57-70	gm
NaCl	-	0.003	"
$MgSO_4$	-	0.0045	"
KNO_3	-	0.0003	"
$FeCl_3$	-	0.0002	(trace)
$CaSO_4$	-	0.03	gm
Distilled water	-	1000	ml

7. Myer's G medium for Blue-green Algae (See Watamabe, 1960):

KNO_3	-	1.0	gm
K_2 HPO $_4$	-	1.0	"
Mg $SO_4 \cdot 7H_2O$	-	0.25	"
Ca $(N0_3)_2 \cdot 4H_2$ 0	-	0.025	"
N a-citrate	-	0.165	"

$FeS0_4 \cdot 7H_2O$	-	0.02	"
A_5 solution	-	1	ml
(pH 8.0)			

8. Fogg's Nitrogen Free Medium (Fogg, 1949) :

KH_2PO_4	-	0.2	gm
$MgSO_4$	-	0.2	"
$CaC1_2$	-	0.1	"
Na_2MO_4	-	0.1	mg
$MgC1_2$	-	0.1	"
H_3B0_3	-	0.1	"
$CuSO_4$	-	0.1	gm
$ZnSO_4$	-	0.1	ml
Fe-EDTA	-	1.0	ml
Distilled water	-	1000	ml

9. Bristol's Solution (Bristol, 1919, 1920, 1923) :

KH_2P0_4	-	1.0	gm
$N aN0_3$	-	1.0	"
$CaC1_2$	-	0.1	"
$MgS0_4$	-	0.3	"
NaCl	-	0.1	"
$FeC1_3$	-	0.01	"
Distilled water	-	1000	ml

Stock solutions of different chemicals were prepared in distilled water. Ferric chloride solution was freshly prepared whenever required. The pH of the culture media was adjusted prior to autoclaving.

Sterilized bottles and flasks were then filled with convenient amount of the medium and were again sterilized in an autoclave at 15 lbs pressure for 20 minutes. After the bottles had cooled down, approximately one gram of the requisite soil was introduced into separate bottles or flasks with the help of a sterile spatula in aseptic conditions. Sterilized culture solution was added to the cultures at suitable intervals to make up the loss of the culture solution partly lost due to evaporation.

Moist cultures were prepared in petridishes which were sterilized in the same way as in the case of bottles. Convenient amount of soil depending on the size of the petridishes were spread evenly in each petridish and moistened with culture medium in the beginning to promote the algal growth. Later on, the petri-dishes were moistened with sterilized distilled water at intervals to prevent drying of the cultures.

All the cultures were placed near open windows facing north where cultures could get only diffused light. From dense growing cultures subcultures were made for the convenience.

Unialgal cultures :- In making sub cultures, a few cells were drawn into a fine pipette, washed in sterilized water and transferred to 1.5 per cent De's modified Beneck's medium with agar in petridishes and test tubes. 1.5 per cent agar medium was prepared by mixing 15 gm of agar in one litre of De's modified Benecks medium. This solution was poured into petri-dishes and test tubes aseptically after autoclaving.

After two weeks, individual algal units were picked up and transferred to fresh agar plates. After 3-4 transfers on agar plates groups of particular algae were transferred to Allen & Arnon's liquid medium for maintenance.

SOIL ANALYSIS

Mechanical composition of soil

Air dried soil samples were gently ground with a wooden pestle. The ground substance was passed through 2 mm diameter sieve. The sieved material was mixed thoroughly and preserved in labelled cloth bags for further analysis. Above soil samples were utilized for physical and chemical analysis.

(a) Soil texture :

Texture of soil was determined by taking known quantity of air dried soil and passing it through different grades of standard sieves for mechanical analysis by using international pipette method (Piper, 1950).

(b) Moisture determination :

It was determined by drying a weighed quantity of soil sample in a weighed aluminium box in an oven at 105°C for 24 hours. The aluminum box along with soil samples were weighed again. Difference in the weight gave the corresponding loss of moisture. The heating was repeated again for 2 to 3 days more till constant loss in weight was obtained. The difference in initial and final weight gave the amount of moisture lost from given soil. From this moisture percentage (for 10 gm) was calculated.

(c) Determination of soil pH :

25 gm of air dried soil were taken in a beaker. 100 ml of boiled water was carefully added until the soil becomes soft enough to allow ready penetration of the electrods. The amount of soil to water was in the ratio of 1:4. This mass was gently

tapped on the table and electrodes were pressed into the soil. Several pH readings were taken on different samples of the same soil. The constant reading was noted down (Piper, 1950).

(d) Determination of organic carbon :

10 gm of soil sample which was finely ground to pass through 0.5 mm sieve, was taken in a 500 ml Erlen Meyer flask. 10 ml of normal potassium dichromate solution was added by means of pipette followed by 20 ml of concentrated sulphuric acid. The contents of the flask were shaked mechanically by for two minutes and set aside on an asbestos pad for exactly half an hour. At the end of the period added 200 ml of water, 10 ml of phosphoric acid and 1 ml of the indicator diphenylamine to the flask. The contents of the flasks attain a dark blue colour. Filtered the contents against 0.5 N ferrous sulphate solution till the contents attain a green colour. This change in colour from blue to green comes too suddenly and as such the end point of the titration followed and located with great care.

Calculations :

Titrate value = Total volume (ml) of dichromate 1 N Solution added - half the volume (ml) of 0.5 N ferrous sulphate solution used in titration

$$\% \text{ of carbon in the soil} = \frac{\text{Titrate value (ml)} \times 0.003 \times 100}{\text{Weight of the soil taken (gm)}}$$

(e) Total nitrogen :

Transferred 10 gm of the air dried soil into a clean dry Kjeldahl flask. Added carefully from the sides of the flask 50 ml of sulphuric acid-salycilic acid mixture followed by 5 gm of sodium thiosulphate and digested the contents of the flask in a fume-cupboard on a low flame till for thing ceases. At this stage added about 10 gm of potassium sulphate and carried out the digestion in a strong flame. Removed the flask from the flame intermittently and shaken the contents carefully in a manner to bring down any stray particle of carbon or undigested material. When the digestion is complete a light blue green colour developed. Continued the digestion for 15 minutes more after the development of the said colour. Removed the flask from the flame and cooled. Just when the flask was slightly warm added carefully 30 ml of water and mixed the contents to prevent caking.

Transferred without loss only a water extract of the digested material in the flask into the distillation flask. Made about 6 extractions using each time about 50 ml of water. Added a few glass beads and 1 or 2 zinc clippings to the distillation flask. Added a mixture of 150 ml of 40 percent NaOH and 5 ml of sodium sulphide to the flask. Ensured that the content of the flask is distinctly alkaline. Carried out the distillation, collecting the distillate into an Erlen mayer flask containing a known excess of 0.1 NH_2S0_4 (50 ml). When about 150-200 ml of the distillate were collected, tested for the absence of ammonia at the tip of the delivery tube. Back titrated the excess of acid with 0.1 N NaOH using methyl red as an indicator.

Calculation :

Titrate value = 0.1 N acid (ml) taken - 0.1 N alkali (ml) used for back titration

$$\% \text{ of nitrogen in soil} = \frac{\text{Titrate value (ml)} \times 0.0014 \times 100}{\text{Weight of the soil taken (gm)}}$$

(f) Available phosphoric acid (P_0O_5) :

Procedure

Transferred into a Winchestor bottle 120 gm of soil and added 1200 ml of 1 percent citric acid solution. Allowed the soil to remain in contact with the solution for 7 days. With occasional shaking, alternately shaken the contents continuously for 24 hours in an end over end shaker. Removed it from the shaker and allowed the contents to settle down and decanted carefully the clear supernatant liquid on siphon off. Passed this extract through a buchner funnel fitted with a dry filter paper and collected the filtrate in a dry flask.

Concentrated 500 ml of the citric acid extract in a large china dish over a water bath, to a small volume of about 30 mI. Transferred the same without loss to a silica basin, evaporated to dryness and ignite to ash. Digested the ash with dilute HCI on a hot water till it is free from chloride. Rejected the residue. Transferred the filtrate to a silica basin, evaporated it to dryness. Extracted the residue with dilute $HN0_3$ and precipitated P_20_5 as ammonium phosphomolybdate and estimated by the volumetric method.

(g) Available potassium

Concentrated 500 ml of the citric acid extract to a small volume (30 ml) and transferred it without loss to a small silica basin (10 cm diameter). Evaporated to dryness, ignited till a brownish red powder is obtained. Carried out ignition over a flame a long period instead of using a strong flame for a some period. Digested it with water over a water bath and filtered through Whatman No. 40 filter paper and collected the filtrate. The filtrate was colourless, if ignition was carried out properly.

Evaporated to dryness in a glassbasin, the colourless filtrate over water bath and precipitated potash as K_2PtCl_6 and estimated as directed by Piper, 1950.

In calculation of K_2O and P_2O_5, it was taken that 500 ml of the citric acid extract corresponds to 50 gm of air dried soil.

Table I:
Table Showing the Districts and Places of Study

Sr. No.	*Districts*							
	Nagpur	*Bhandara*	*Chandrapur*	*Wardha*	*Amravati*	*Yavatmal*	*Buldana*	*Akola*
1.	Nagpur	Bhandara	Chimur	Wardha	Amaravati	Yavatmal	Buldana	Akola
2.	Kamptee	Tumsar	Warora	Pulgaon	Badnera	Pusad	Khamgaon	Akot
3.	Ramtek	Tiroda	Sironcha	Hinganghat	Dhamangaon	Kalamba	Malkapur	Murtizapur
4.	Katol	Gondia	Mool	Arvi	Morshi	Wani	Jalgaon–Jamod	Patur
5.	Saoner	Amgaon	Gadchiroli	Karanja		Khandala		Nalgaon
6.	Kalmeshwar	Salekasa	Sindewahi					Umari
7.	Kuhi	Lakhni	Nagbhir					
8.	Umrer	Sakoli	Brahmapuri					
9.	Butibori	Arjuni	Armori					

MAP – IV

TOPOGRAPHY OF NAGPUR DISTRICT SHOWING THE PLACES OF STUDY

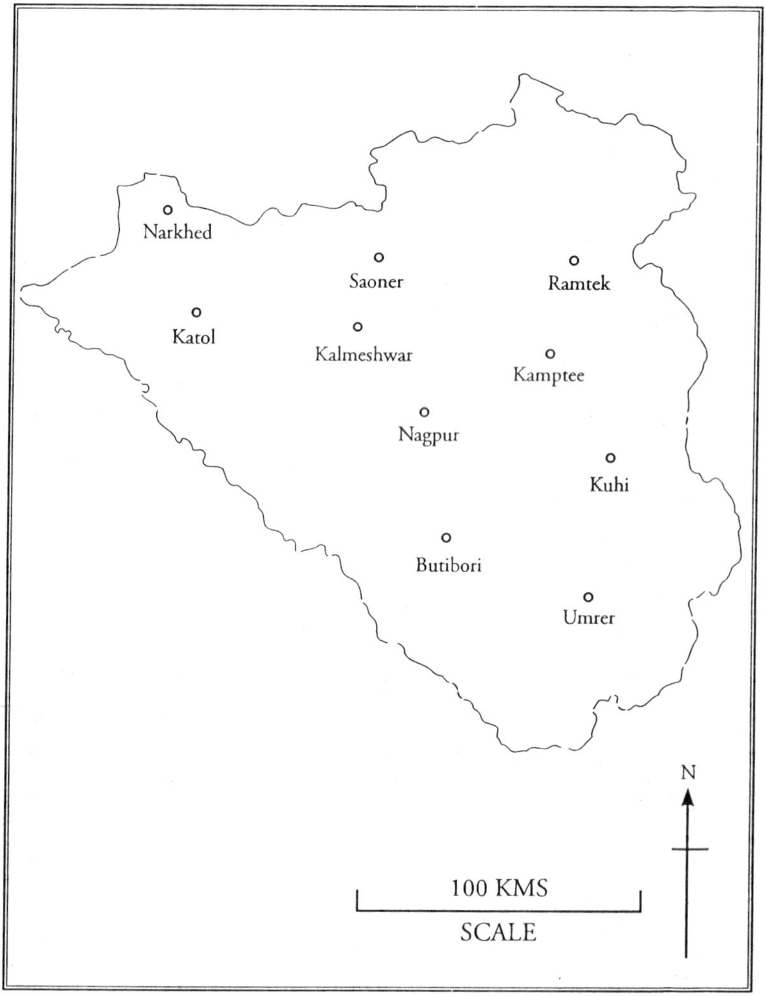

MAP – V

TOPOGRAPHY OF BHANDARA DISTRICT SHOWING THE PLACES OF STUDY

MAP – VI

TOPOGRAPHY OF CHANDRAPUR DISTRICT SHOWING THE PLACES OF STUDY

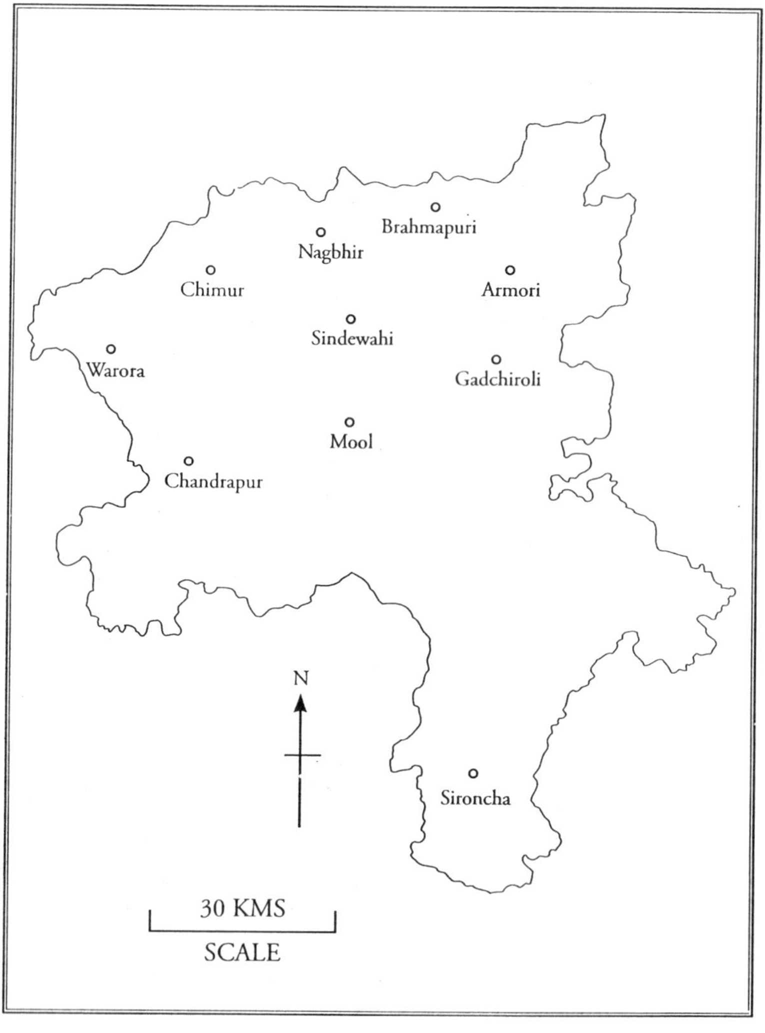

MAP – VII

TOPOGRAPHY OF WARDHA DISTRICT SHOWING THE PLACES OF STUDY

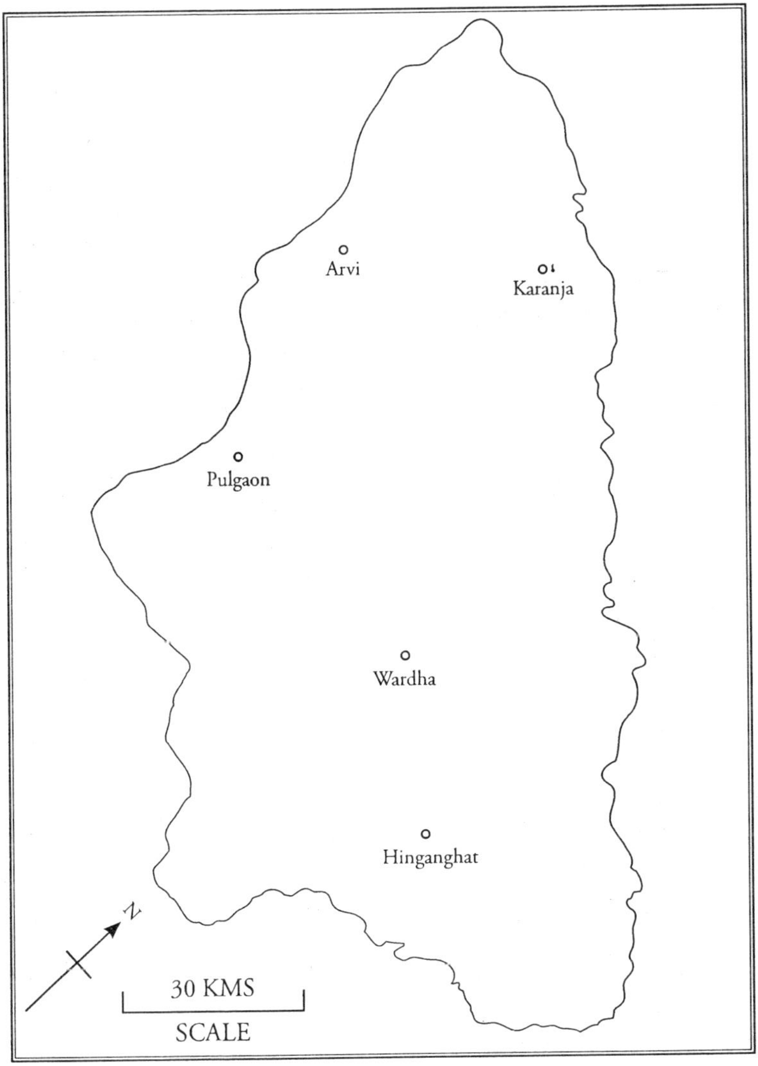

MAP – VIII

TOPOGRAPHY OF AMARAVATI DISTRICT SHOWING THE PLACES OF STUDY

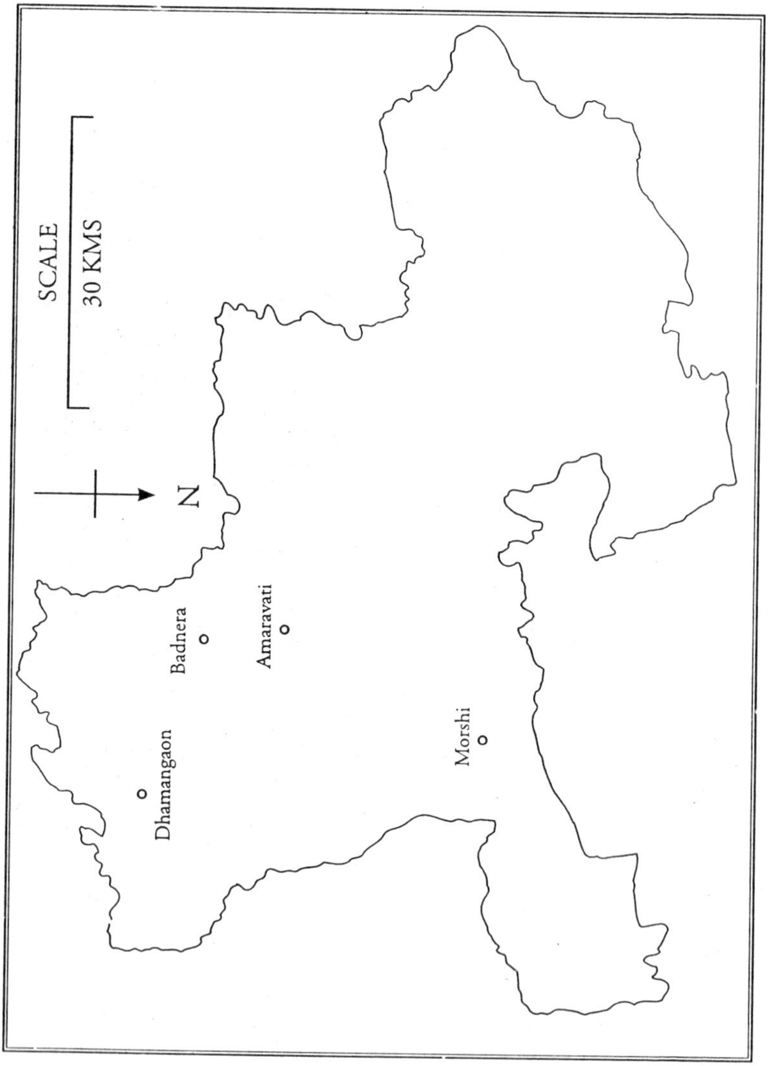

MAP – IX

TOPOGRAPHY OF YAVATMAL DISTRICT SHOWING THE PLACES OF STUDY

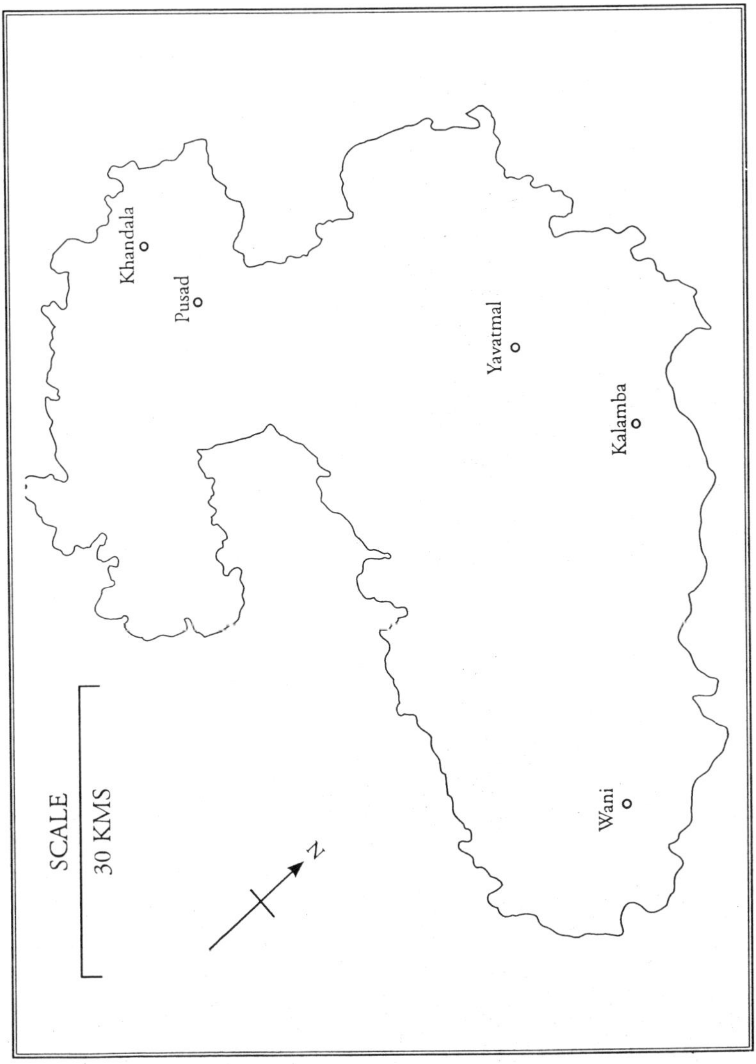

MAP – X

TOPOGRAPHY OF BULDANA DISTRICT SHOWING THE PLACES OF STUDY

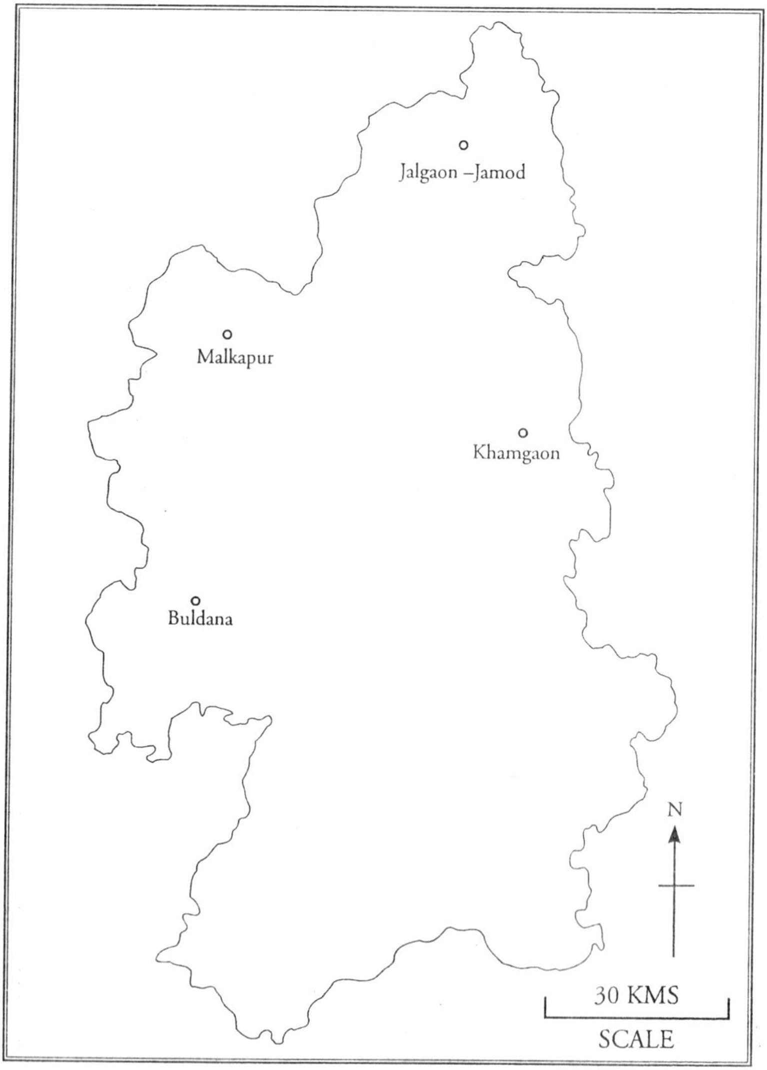

MAP – XI

TOPOGRAPHY OF AKOLA DISTRICT SHOWING THE PLACES OF STUDY

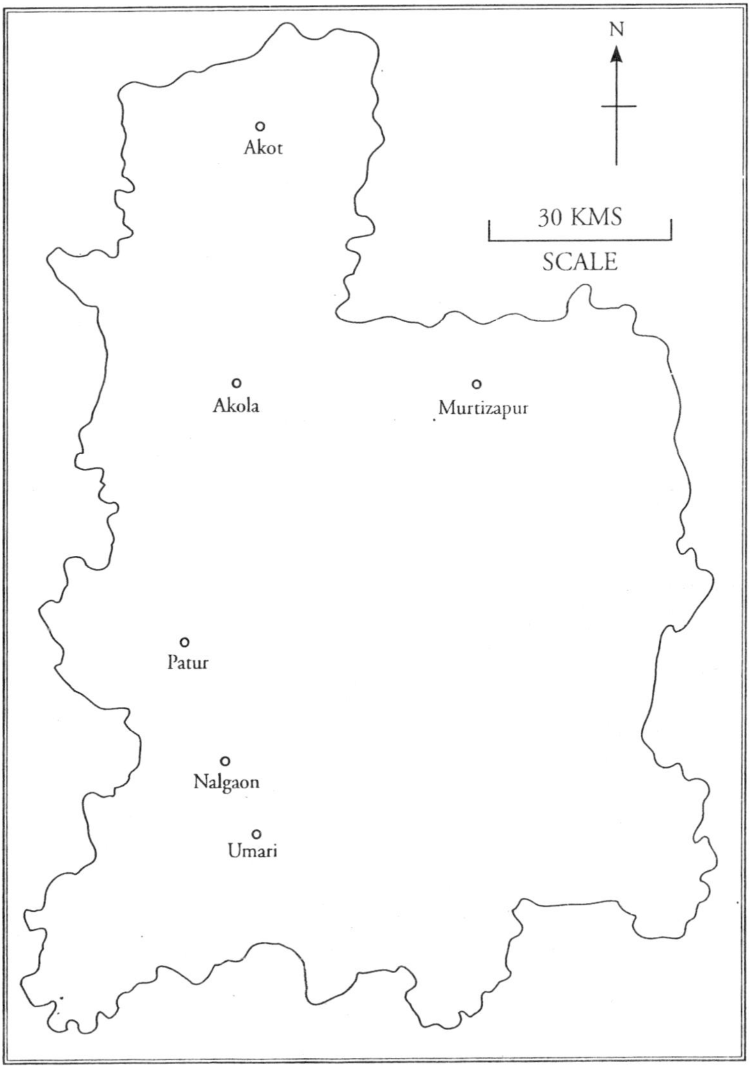

Chapter 3

RESULTS AND DISCUSSION

The algal flora of paddy field of Vidarbha region were worked out in this part of the present work. To study the different aspects of algae of paddy fields, the algal flora present in this field must be known. The list of algal flora of a place is confirmed by the culture as well as fresh collection studies. Thus the elimination of a particular species is minimized. To correlate the habitat of a particular algae, soil analysis of the region as well as its state of occurrence is also observed.

Vidarbha comprises 8 districts and out of this 8 only in Nagpur, Bhandara and Chandrapur, the main crop is paddy. In rest of the districts, even though the main crop is not paddy, still scattered paddy fields could be noticed. From each district collections of algae and culture of soils were made.

1. Paddy field algal flora of Nagpur district:

Nagpur district lies between 20° 35' and 21° 44' North latitude and 78° 45' and 79° 35' east longitude, it has an area of 9930.06 kms. The total area under forest is 0.52 lac hectares which is 6% of the geographical area of the district. In all 62% of the total area i.e. 5.33 lac hectares were under cultivation. The main kharif crops are Jowar, Tur, Cotton and Rice. The average rainfall of the district is 1172 mm most of which is received during the months June to September. The temperature range in the district is from 13°C during winter to 45°C during summer.

The results of physical and chemical analysis of Nagpur district are given in Table no.IV. Although the soils in this district are properly called black cotton soils, they are not as deep as normal black cotton soils in India. The district has different types of soil structures ranging from brown black calcareous soils in the western parts, to yellowish brown and red soils of mixed parent material in the eastern parts. Black soils are formed from hasaltic rocks and yellow soils are mostly derived from alluvium and sedimentary rocks. The soils are alkaline in reaction, and the total soluble salts are also not high. They are clacy in texture and have fair amount of calcium carbonate. The analyzed soil samples were well supplied with nitrogen available phosphate and available potash.

Total 9 places of the district were studied. From each place 5 to 10 soil samples were worked out. In all 211 algal taxa could be identified from the studies. These taxa belonged 152 to Cyanophyta, 35 to Chlorophyta, 3 to Xanthophyta, 5 to Euglenophyta and 16 to Bacillariophyta. The total algal taxa comprises *Microcystis* (4), *Chroococcus* (9), *Gleocapsa* (1), *Gleothece* (4), *Aphanocapsa* (6) *Aphanothece* (6), *Synechococcus* (1), *Synechocystis* (1), *Merismopedia* (1), *Chlorogloea* (1), *Arthospira* (2), *Oscillatoria* (24), *Phormidium* (14), *Lyngbya* (17), *Schizothrin* (1), *Symploca* (2), *Microcoleus* (2), *Anabaenopsis* (1), *Cylinderospermum* (2) *Nostoc* (10), *Anabaena*(11), *Pseudanabena* (2), *Aulosira* (2), *Plectonema* (2), *Camplyl onemopsis* (1),

Scytonematopsis (1),*Scytonema* (3),*Tolypothrix* (3), *Microchaete* (3),*Fortiea* (1), *Calothrix* (7), *Hapalesihon* (3), *Fischrella*(1), *Stigonema* (2), *Chlamydomonas* (2), *Chlorococcum* (2), *Trebouria* (1), *Tetraedron*(1), *Ankistrodesmus* (1), *Scenedesmus*(4), *Ulothrix*(6), *Geminella* (2), *Protococcus* (1), *Chaetophora* (1), *Balbochaete* (1), *Oedocladium* (1), *Oedogonium*(4), *Spirogyra* (1), *Sirocladium*(3), *Sirogonium*(1), *Closterium*(4), *Cosmarium* (1), *Vaucheria* (2), *Botrydium*(1), *Euglena* (2), *Phacus* (2), *Trachelomonas* (1), *Cyclotella* (1), *Fragillaria*(1), *Synedra*(1), *Achnanthus* (1), *Cymbella* (3) *Nitzschia* (5).

The algal taxa with higher frequency of occurrence i.e. above 50% from the places studied were *Gleothece palea, Chlamydomonasglobosa, Chlamydomonasmucicola, Chlorococcum humicolo, Scenedesmus dimorphus, Ulothrix variabilis, Protococcus viridis, Closterium parvulum, Cosmarium granatum, Fraqillaria brevistriata F. elonqata, Syndera affinis, Achnanthes delicatula, Cymbella austriaca, Cymbella cymbiformis, Nitzschia,angustata* and *Nitzschia dissipata.*

2. Paddy field algal flora of Bhandara district:

In all 9 places of Bhandara district were studied to make the paddy field algal flora of the said district. Bhandara district occupies an area of about 9383.57 sq km between latitudes 20° 20' and 20° 36' and longitude 79° 30' and 80° 40'. It is situated in the North-eastern corner of Maharashtra State. The total area under forest is 2.63 lac hectares which constitutes 24 percent of the total area. The area under cultivation is around 4.71 lac hectares, which accounts for 50 per cent of the geographical area. There are a number of irrigation tanks, which is a distinctive feature of the district. The main crop of the district is paddy but wheat is also cultivated in some areas. The average annual rain fall is about 1451.15 mm almost throughout the district. The temperature varies between 15°C during January to 43°C during May.

The result of physical and chemical soil analysis of Bhandara district are given in Table No.V. The main soil types of the district are 'Kali', 'Khanhan', 'Sihar', 'Morand', 'Khardi' and 'Bardi'. Kali and Khanhan types of soils are not much under cultivation. The most important soils are Morand and Sihar, which together occupy the major portion of the district. Of these, Sihar is a reddish yellow soil, whereas Morand is a light coloured loam, containing sand and stones. Most of these soils are alkaline in reaction and sand in texture.

The list of algal taxa observed from Bhandara district were given along with other districts in Table II. In all 132 taxa of algae could be identified from the paddy fields of Bhandara district. The total alga taxa comprises of 62 Cyanophycean taxa, 34 Chlorophycean taxa, 10 Euglenophycean taxa and 26 Bacillariophycean taxa. The Cyanophycean genera with its number of species were *Microcystis* (6) *chroococcus* (3), *Aphanocapsa* (2), *Spirulina* (3) *Oscillatoria* (9), *Phormidium* (3), *Lyngbya* (5), *Schizothrix* (1),*Symploca* (1), *Hydrocoleus* (1), *Cylinderospermum* (2), *Nostoc* (4), *Anabaena* (7), *Camptylonemopsis* (1), *Scytonematopsis* (1), *Scytonema* (2), *Tolyopothrix* (2), *Microchaete* (2), *Calothrix* (4), *Hapalosiphon* (2) and *Stigonema* (1). The *Chlorophycean* genera were *Chlamydomonas* (2), *Chlorococcum* (1) *Oocystis* (3), *Scenodesmus* (2), *Ulothrix* (3), *Protococcus* (1), *Chaetophora* (2), *Coleochaete* (1), *Bulbochaete* (1), *Oedocladium* (1), *Oedogonium* (5), *Spirogyra* (8), *Closterium* (2) and *Cosmarium* (2). The Euglenophycean genera were *Euglena* (2), *Phacus* (4), *Trachelomonas* (4). The Bacillariophycean genera were *Fragillaria* (1), *Synedra* (1),

Achnanthus (3), *Navicula* (9), *Cymbella* (3), *Rhopalodia* (1), *Gomphonema* (2), *Hantzschia* (1), *Nitzschia* (3).

The algal taxa with higher frequency of occurrence (above 50 %) from the places studied were *Microcystis, holosatica, Microcystis protocystis, Chroococcus macrococcus, Chroococcus turgidus, Aphanocapsa biformis, Aphanocapsa elachista, Oscillatoria amoena, Oscillatoria curviceps, O.curviceps v. angusta, decolorata, O. princeps, O. salina, Phormidium foveolarum, P. jenkelianum, Lyngbya arugneo-coerulea, Schizothrix tenuis, Cylinderospermum mucicola v. longispora, Nostoc commune, N.calcicola, N.microscopica, N. palludosum, Anabaena anomala, A. laxa, A. naviculoides, A. torulosa, A. variabilis, A. variabilis v. kashiensis, Camptylonemopsis iyengarii, Scytonematopsis woronchinii, Sctonema fremyii, Tolypothrix bouteillei, T. byssoidea, Microchaete calothrichoides, M. tenera, Calothrix brevissima, v. moniliforme, C.epiphytica, Hapalosiphon intricatus, H. welwitschii, Chlamydomonas globosa, C. mucicola, Chlorococcum humicolo, Scenedesmus dimorphus, Ulothrix variabilis, Protococcus viridis, Spirogyra biformis, S. dictiospora, Closterium acerosum, C. parvulum, Cosmarium granatum, Fragillaria brevistrata, F. elongata, Synedra affinis, Achnanthes delicatula, Navicula clavata, Navicula grivilleii, Pinnularia viridis, Cymbella austriaca, Cymbella cymbiformis, Nitzchia dissipata, Nitzschia gracillis.*

3. Paddy field algal flora of Chandrapur district:

Chandrapur district lies in the eastern corner of Maharashtra. It is border district with Andhra Pradesh to its south and Madhya Pradesh to its east. The district is located between latitudes 18^0 40' and 20^0 40' and longitudes 78^0 44' and 83^0 50' and occupies an area of about 2418.08 square km. The total area under forest is 14.36 lac hectares and this constitutes 60% of the total area. The total cultivable area is 5.42 lac hectares, which is hardly 22% of the area of district. The main crops of the district are paddy, jowar, cotton, wheat and gram. Paddy occupies 34% of the cultivable area. The district receives an average rainfall of 1406.40 mm mostly of which is received during the months June to September. It varies from 1168 to 1524 mm increasing from west to east. The temperature ranges between 10^{0}C in December and 45^0 C in May.

The result of physical and chemical soil analysis of this district are given in Table no. VI. The soils in the district range from deep heavy to light soils. 'Morand' soils, which are most common in the district, are light coloured loams and occasionally contain particles of stones. The soils are very slightly alkaline in reaction, sandy to loamy in texture and contain very little calcium carbonate. They are low in nitrogen content.

The list of algal taxa with their places of occurrence were given in Table No.II. In all 137 taxa of algae could be identified from this district; out of which 87 Cyanophycean taxa, 25 Chlorophycean taxa, 1 Xantophycean taxa, 2 Euglenophycean taxa and 22 Bacillariophycean taxa. The Cyanophycean genera with its number of species were *Chroococcus* (6), *Gleocapsa* (3), *Aphanocapsa* (3) *Spirulina* (1), *Oscillatoria* (15), *Phormidium* (5), *Lyngbya* (7), *Schizothrix* (1), *Symploca* (1), *Hydrocoleus* (1), *Cylinderospermum* (2), *Nostoc* (7), *Anabaena* (9), *Aulosira* (1), *Plectonema* (1), *Camptylonemopsis* (1), *Scytonematopsis* (1), *Scytonema* (3), *Tolypothrix* (2), *Microchaete* (2), *Fortiea* (1), *Calothrix* (5), *Gloeotrichia* (3), *Hapalosiphon* (2) *Fischerella*.(l), *Stigonema* (1). The Chlorophycean genera were Chlamydomonas (1), *Cartaria* (1), *Characium* (1) *Chlorococcuum* (1), *Trebouxia* (1), *Chlorella* (1), *Tetradron* (1), *Trochiscia* (1), *Oocystis* (2), *Ankistrodesmus* (1), *Selenastrum* (1), *Scenedesmus* (1),

Ulothrix (2), *Geminella* (1), *Protococcus* (1), *Chaetophora* (1), *Oedogonium* (2), *Spirogyra* (2), *Closterium* (3), *Cosmarium* (2). Xanthophycean genera was *Vaucheria* (1). The Euglenophycean genera were *Trachelomonas* (2), The Bacillariophycean genera were *Cyclotella* (1), *Fragillaria* (1), *Synedra* (1), *Achnanthes* (1), *Gyrosigma* (1), *Navicula* (8), *Pinnularia* (1), *Amphora* (2), *Cymbella* (3), *Nitzschia* (3).

The algal taxa with higher frequencies of occurrence (above 50%) from the places studied were *Chroococcus* macrococcus, *Chroococcus spelaeus, C. turgidus, Aphanocapsa biformis. A. fonticola. Oscillatoria amoena,* O. *curviceps, O. curviceps V. angusta,. O. decolorata, O. princeps, O.salina, Phormidium foveolarum, P. jenkelianum. Lyngbya aerugineo-coerulea, Schizothrix tenuis, Cylinderospermum mucicola, v. longispora, Nostoc commune, N.calcicola, N. microscopica, N. palludosum, N. spongiaeforme, Anabaena doliolum, A. naviculoides, A. randhawae, A. variabilis, A. variabilis* var, *Kashiensis, Camtylonemopais iyengarii, Scytonematopsis woronichinii, Scytonema cincinnatus, S. millei, Tolypothrix bouteillei, T. byssoides, Microchaete calothrichoides, M. tenera, Calothrix brevissima, V. moniliforme, C. epiphytica, C. membrancea, C. marchica, v. intermedia, Hapalosiphon intricatus, H. welwitschii, Stigonema harmoides, Chlamydomonas globosa, Chlorococcum humicolo, Scenedesmus dimorphus, Ulothrix variabilis, Protococcus viridis, Closterium acerosum, C. parvulum, Cosmarium granatum, C. trilobatum, Fragillaria brevistriata, f. elongata, Synedra affinis, Navicula clavata, N.gravillei, Pinnularia microstauron, Cymbella austriaca, C.cymbiformis, Nitzschia dissipata, N. gracillis.*

4. Paddy field Algal flora of Wardha district:

Wardha district occupies an area of 6304.06 sq. km. The districts located between 20° 18' and 21° 21' North latitude and 78° 4' and 79° 15' East longitudes. The area under forest is 0.66 lac hectares i.e. 10% of the total area and the total cultivable land is 4.06 lac hectares, which is 65 % of the total area. The main kharif crops are Jowar and Cotton. Wheat is also grown there to a small extent. Paddy are grown very rarely. The average rainfall of the district is 1101.94 mm mostly precipitating between June and September. It varies from 948.9 to 1125.22 mm in different talukas. The temperature minimum and maximum are 15°C and 45°C respectively.

The result of physical and chemical soil analysis of this district are given in Table No.VII. Black cotton soils containing varying amounts of lime are the most predominant type. The soils contain good amounts of all plant nutrients. They are moderately alkaline, clayey and low in salt content.

In all 182 taxa of algae could be identified from 5 places of Wardha of the total algal taxa, 135 belongs to Cyanophyta, 21 belongs to Chlorophyta, 1 to Xanthophyta, 7 to Euglenophyta and 18 belongs to Bacillariophyta. The total algal taxa comprises *Microcystis* (3) *Chroococcus* (8), *Gleocapsa* (1), *Gleothece* (2), *Aphanocapsa* (6), *Aphanothece* (1), *Synechococcus* (3), *Merismopedia* (1) *Chlorogloea* (1), *Chroococcidiopsis* (1), *Dermocarpa* (1) *Arthospira* (2), *Spirulina* (2), *Oscillatoria* (24), *Phormidium* (12), *Lyngbya* (15), *Schizothrix* (1), *Symploca* (2), *Microcoleus*(1), *Cylinderospermum*(2),*Nostoc*(10),*Anabaena*(9), *Pseudnabaena* (2), *Aulosira* (1), *lectonema* (1), *Camptylonemopsis* (1), *Scytonematopsis* (1), *Scytonema* (4), *Tolypothrix* (2), *Microchaete* (4), *Calothrix* (5), *Gloeotrichia* (1), *Hapalosiphon* (2), *Westiellopsis* (1), *Stigonema* (1), *Chlamydomonos* (2), *Chlorococcum* (2), *Chlorella* (1), *Oocystis* (1), *Scenedesmus* (3), *Ulothrix* (2), *Rhizoclonium* (1), *Protococcus* (1), *Coleochaete* (1), *Oedocladium* (1), *Sirogonium* (1), *Closterium* (2), *Cosmarium* (3), *Botrydium* (1), *Euglena* (4), *Phacus* (1), *Trachelomona*

(2), *Cyclotella* (1), *Fragillaria* (1), *Synedra* (1), *Achnanthes* (1), *Navicula* (6), *Pinnularia* (1), *Amphora* (1), *Cymbella* (2), and *Nitzschia* (3).

The algal taxa with higher frequency of occurrence (above 50%) from the places studied were *Chroococcus granulosus, C. macrococcs, C. spelasus, C. turgidus, Aphanocapsa biformis, A. fonticola, Arthospira jenneri, Oscillatoria amoena, O. curviceps, O. curviceps v. angusta, O. decolorata, O. princeps, O. salina, Phormidium foveolarum, P. jenkelianum, Lyngbya cryptovaginata, L. ceylanica, Symloca cartilaginea, Cylindrospermum mucicola* var.*lonispora, Nostoc commune, N. calcicola, N. microscopicum, N. paludosum, Anabaena anomola, A. naviculoides, A. torulosa, A. variabilis, A. variabilis V. kashiensis, Camptylonemopsis iyengarii, Scytonematopsis woronichinii, Scytonema fremyii, Tolypothrix bouteillei, T. byssoidea, Microchaete calothrichoides, M. tenera, Calothrix brevissima var. moniliforme, C. epiphytica, C. membranacea, C. marchica v. intermedia, Hapalosiphon intricatus, H. welwitschii, Stigonema hormoides, Chlamydomonas globosa, Chlorococcum humicolo, Scenedesmu dimorhus, Ulothrix variabilis, Protococcus viridis, Closterium acerosum, C. parvulum, cosmarium granatum, Fragillaria brevistriata f. elongata Synedra affinis, Achnanthes delicatula. Navicula aponina, N. clavata, N. gravillei, N. lancolata, N. tennelloides, Pinnularia microstauron, Amphora ovalis, Cymbella austriaca, C. cymbiformis, Nitzschia dissipata , N. gracillia, N. palea.*

5. Paddy field algal flora of Amravati district:

Amravati district lies between 20° 15' and 21° 45' North latitude and 76° 30' and 78°15' east longitude. The total area of the district is 12149.6 sq. kilometers; out of which 3.51 lac hectares constitutes forest and 6.70 lac hectares land is under cultivation. Forest constitutes 29% and cultivated land constitutes 55% of the total area of the district. The main crops cultivated are kharif jowar, cotton and ground nut. The mean average annual rainfall of the district is 923.8 mm which varies between 711.2 and 1171.6 mm in different talukas of the district. The temperature minimum and maximum are 10° C and 45° C respectively.

The result of physical and soil analysis of this district is given in Table VIII. Brown black calcareous soils are the predominant type of the district. These soils are formed from basaltic rocks. The soils are generally alkaline in reaction and clayey in texture. List of algal taxa with their places of occurrence were given in table II. In all 185 taxa of algae could be identified from 4 places of the district. This taxa were belonged 134 to Cyanophyta, 26 to Chlorophyta, 1 to Xanthophyta, 6 to Euglenophyta and 18 to Bacillariophyta. The total algal taxa comprises *Microcystis* (2), *Chrococcus* (9), *Gleocapsa* (5), *Gloeothece* (2), *Aphanocapsa* (4) *Aphanothece* (4), *Synechococcus* (2), *Synechocystis* (1) *Merismopedia* (1), *Arthoseira* (2), *Oscillatoria* (22), *Phormidium* (14), *Lyngbya* (12), *Schizothrix* (1), *Symploca* (3), *Anabaenopsis* (1), *Cylinderospermum* (5), *Nostoc* (11), *Anabaena* (12), *Aulosira* (2), *Plectonema* (1), *Camptylonemopsis* (l), *Scytonematopsis* (1), *Scytonema* (2), *Tolypothrix* (2), *Microchaete* (2), *Fortea* (1), *Calothrix* (6), *Hapalosiphon* (2), *Stigonema* (1), *Chlamydomonas* (2), *Chlorococcum* (2), *Tetraedron* (1), *Scenedesmus* (3), *Ulothrix* (2), *Microspora* (1), *Protococcus* (1), *Oedogonium* (1), *Spirogyra* (5), *Sirocladium* (1), *Closterium* (5), *Cosmarium* (2), *Vaucheria* (1), *Euglena* (2), *Phacus* (3), *Cyclotella* (1) *fragillaria* (1), *Synedra* (2), *Achnanthes* (1), *Navicula* (4) *Amphora* (2), *Cymbella* (4), *Nitzschia* (3).

The algal taxa with higher frequency of occurrence (above 50%) from the places studied are *Chroococcus macrococcus,*

C. spelaeus, C. turqidus, Aphanocapsa biformis, A. fonticola, Oscillatoria amoena, O. curviceps, O. curviceps v. anqusta, O. decolorata, O. princeps, O. salina, Phormidium foveolarum, P.jenkelianum, Lyngbya aeruqinea-coerulea, L.shackletoni, Cylindrospermum mucicola v. longispora, Nostoc commune, N. calcicola, N. microscopicum, N. paludosum, Anabaena amomala, A. laxa, A. naviculoides, A. torulosa, A. variabilis, A. variabilis v. kashiensis, Camptylonemopsis iyengarii, Scytonematopsis woronichinii, Scytonema fremyii, Tolypothrix bouteillei, T. byssoidea, Microchaete calothrichoides, M. tenera, Calothrix brevissima var. moniliforme, C. epiphytica, C. membrancea, c. marchica, v. intermedia, Hapalosiphon intricatus, H. welwitschii, Stignonema hormoides, Chlamydomonas globosa, Chlorococcum humicolo, Scenedesmus dimorphus, Ulothrix variabilis, Protococcus viridis, Closterium acerosum, C. parvulum, Cosmarium granatum, Fragillaria brevistriata f. elongata, Synedra affinis, Achnantaece delicatula, Navicula clavata, N. gravillei, Pinnularia microstauron, Cymbella austriaca, C. cymbiformis, Nitzschia dissipata, N. gracillia.

6. Paddy field algal flora of Yavatmal district:

Yavatmal district occupies an area of about 23,925 square kilometers between north latitudes 19° 28' and 20° 48' and east longitudes 77° 19' and 79° 7'. It is located in the eastern part of Maharashtra on the boarder of Andhra Pradesh. The average annual rainfall of about 986.44 mm which occurs mostly between the months of June to September. The temperature minima and maxima are 15^{0}C and 43° C respectively. The total area under forest is 2.98 lac hectares, and this constitutes 33% of the total area. The area under cultivation is a little more than 7.29 lac hectares which corresponds to about 54 % of the total area of the district. The most important crops are kharif jowar, cotton, pulses and ground nut which together occupy more than 90% of the cultivated area. The district is, therefore, a predominantly kharif cropped area.

The results of soil analysis were given in table No.IX. The soils which are derived from Deccan trap vary in their characteristics according to their location. The medium black soil is the predominant type occurs extensively in different parts of the districts. The soils are alkaline in reaction and clayey loam in texture. They contain good amounts of plant nutrients.

List of algal taxa with their places of occurrence were given in table II. In all 167 taxa of algae could be identified from 5 places of the district. The taxa were belonged 28 to Cyanophyta, 22 to Chlorophyta, 3 to Euglenophyta, and 14 to Bacillariophyta. The total taxa includes *Microcystis* (3), *Chroococcus* (7), *Gloeocapsa* (3), *Gloeothece* (3), *Aphanocapsa* (4), *Aphanothece* (2), *Synechocystis* (1), *Merismopedia* (1), *Chlorogloea* (1) *Spirulina* (1), *Ocillatoria* (24), *Phormidium* (16), *Lyngbya* (9), *Schizothrix* (2), *Symploca* (1), *Microcoleus* (2), *Anabaenopsis* (1), *Cylinderopermum* (1), *Nostoc* (4), *Anabaena* (11), *Pseudanabaena* (2), *Aulosira* (2), *Camtylonemopsis* (1), *Scytonematopsis* (1), *Scytonema* (3), *Tolypothrix* (2), *Microchaete* (5), *Calothrix* (6), *Dichothrix* (1), *Rivularia* (1) , *Nostochopsis* (1), *Hapalosiphon* (3), *Westiellopsis* (1), *Stignonema* (2), *Chlamydomonas* (1), *Chlorococcum* (1), *Chlorella* (2), *Tetraedrom* (1), *Trochiacia* (2), *Scenedesmus* (3), *Ulothrix* (1), *Rhizoclonium* (1), *Protococcus* (1), *Oedogonium* (2), *Spirogyra* (3), *Closterium* (3), *Cosmarium* (1), *Euglena* (2), *Phacus* (1), *Fragillaria* (1) *Synedra* (1), *Achnanthes* (1), *Navicula* (4), *Pinnularia* (1), *Cymbella* (2), *Hantzchia* (1), *Nitzschia* (3).

The majority of the algae (above 50%) is present in *Microcystis insigonis*, *Chroococcus macrococcus*, *Chroococcus minutus*, *C.*

spelaeus, C. turoidus, C. tenax, Gloeocapsa stegophila, Gloeothece rupestris, Aphanocapsa biformis, A. fonticola, Oscillatoria amoena, O. curviceps, curviceps v. *angusta, O. decolorata, O. princeps, O. salina, Phormidium foveolarum, P. jenkelianum, Lyngbya cryptovaginata, Schizothrix tenuis, Cylindrospermum musicola var. longispora, Nostoc commune, N. calcicola, N. microscopicum, N. paludosum, Anabaena anomala, A. laxa, A. naviculoides, A. torulosa, A. variabilis, A. variabilis v. kashiensis Camptylonemopsis, iyengarii, Scytonemotopsis, woronichinii Scytonema fremyii, Tolypothrix bouteillei, T. byssoidea, Microchaete aequalis, calothrichoides, M. tenera, Calothrix brevissima var. moniliforme, C. epiphytica, C. membranace, C. marchica, C. intermedia, Hapalosiphon intricatus, H. welwitschii, Stigonema hormoides, Chlamydomonas globosa, Chlorococcum humicolo, Chorella vulqaris, Scenedesmus dimorphus, Ulothrix variabilis, Protococcus viridis, Closterium acerosum, C. parvulum, Cosmarium granatum, Fragillaria brevistriata, f. elongata, Synedra affinis, Achnanthes delicatula, Navicula clavata, N. gravillei, N. lanceolata, N. major, Pinnularia microstauron, Cymbella austriaca, C. cymbiformis, Nantzschia anphioxya, Nitzschia gracillis, Nitzschia palea.*

7. Paddy field algal flora of Buldana district:

Buldana district lies between 19^0 51' and 21^0 17' north latitude and 75^0 75' and 76^{0}49' East longitude. It has an area of 9702 sq km. The total area under forest is 1.22 lac hectares which forms 12 per cent of the total geographical area of the district. The total area under cultivation is 6.91 lac hectares and this corresponds to 70 per cent of the area of the district. The main crops are kharif jowar, pulses and cotton, which together account for 85 per cent of the cropped area. The rest of the area is under wheat and other crops.

The annual average rainfall is 799.08 mm which is received during the period June to August. The rainfall in different talukas of the district varies from 711.2 - 914.4 mm. Minimum temperature in the district was 15^{0} C and maximum 44^{0}C.

The result of physical and chemical analysis of soils of Buldana district were given in table no.X. Black cotton soils are the predominent type of the district. They vary in depth and colour according to their topographical location. The soils in the northern plains are mostly deep black loams. The soils of the southern plateau are medium deep and contain fair amounts of lime. Those on the higher elevations are reddish brown and shallow with 'murum' layer below. On higher elevations, coarse soils locally known as 'Barad' or 'Murmad' are noticed. Some patches of saline lands are also present here and there.

The total algal taxa with their place of occurrence were given in Table No.II. In all 193 algal taxa could be identified from 4 different places of Buldana district. Of these 193 taxa, 122 belonged to Cyanophyta, 38 belonged to Chlorophyta, 13 belonged to Euglenophyta and 20 belonged to Bacillariophyta.

The total algal taxa were comprises *Microcystis* (2), *Chroococcus* (7), *Gleocapsa* (3), *Aphanocapsa* (5) *Aphanothece* (), *Synechococcus* (1), *Synechocystis* (1), *Merismopedia* (1), *Chlorogloea* (1), *Chroococcidiopsis* (1), *Democarpa* (1), *Arthospira* (1), *Spirulina* (1), *Oscillatoria* (28), *Phormidium* (14) *Lyngbya* (7), *Schizothrix* (2), *Microcoleus* (1), *Anabaenopsis* (1), *Cylindrospermum* (2), *Nostoc* (S), *Anabaena* (8), *Pseudanabaena* (1), *Nidularia* (1) *Plectonema* (1), *Camptylonemopsis* (1), *Scytonematopsis* (1), *Scytonema* (1), *Tolypothrix* (3), *Microchaete* (4), *Calothrix* (6), *Rivularia* (1), *Hapalosiphon* (2), *Stigonema* (1), *Chlamydomonas* (3), *Chlorococcum* (2), *Trebouxia* (1), *Chlorella* (1), *Protosiphon* (1), *Scenedesmus* (4), *Ulothrix* (4), *Geminella* (2), *Microspora*

(1), *Protococcus* (1), *Oedogonium* (S), *Spirogyra* (4), *Sirocladium* (1), *Closterium* (3), *Cosmarium* (5), *Euglena* (4), *Phacus* (4), *Trachelomonas* (5), *Fragillaria* (1), *Cyclotella* (1) *Synedra* (2), *Achnanthes* (2), *Stauronis* (1), *Navicula* (5), *Pinnularia* (2), *Cymbella* (3), *Nitzschia* (2), *Surirella* (1).

The algal taxa which occurred with the high frequency (above 50%) in this district were *Chroococcus micrococcus, C. spelaeus, C. turqidus, Aphanocapsa biformis, A. fonticola, Oscillatoria amoena, O. curviceps, O. curviceps v. angusta, O. declarata, O. princeps, O. salina, Phormidium foveolarum, P. jenkelianum, Lyngbya aeruqineo-coerulea, L. gardneri, L. kuetzingiana, L. martensiana, Schizothrix tenuis, Anabaenopsis tanganyikae, Cylindrospermum majus, C. musicola Var. longispora, Nostoc commune, N. calcicola, N. entophytum, N. microscopicum, N. paludosum, Anabaena anomala, A. laxa, A. naviculoides, A. torulosa, A. variabilis, A. variabilis v. kashiensis, Camptylonemopsis, iyengarii, Scytonematopsis woronichinii, Scytonema fremyii Tolypothrix bouteillei, T. byssoidea, T. nodosa, Microchaete calothrichoides, M.tenera, Calothrix braunii, C. brevissima var moniliforme. C. epiphytica, C.membranacea, C.marchica V. intermedia, Hapalosiphon intricatus, Hapalosiphon welwitschii, Stigonema hormoides, Chlamydomonas globosa, Chlorococcum humicolo, Scenedesmus dimorphus, Ulothrix variabilis, Protococcus viridis, Closterium acerosum, C.parvulum, Cosmarium granatum, Fragillaria brevistriata F. elongata., Synedra affinis, Achnanthes delicatula, Navicula clavata, N. grivillei, Pinnlaria microstauron, Cymbella austriaca, C. cymbiformi, Nitzschia gracillia.*

8. Paddy field algal flora of Akola District:

Akola district, with an area of 10,595 square kilometers, lies between 19^0 51' and $21^0$16' North latitude and $76^0$38' and $77^0$44' East longitude. The annual average rainfall of the district is 833.37 mm, most of which is received during the period of June to August. It varies from 736.6 to 899.0 mm in different talukas. The temperature in the district ranged from 13°C - 45°C.

The total area under forest is 0.87 lac hectares which is about 8 per cent of the total geographical area of the district. The total area under cultivation is 7.71 lac hectares and this constitutes 73 per cent of the total area of the district. The main crops are jowar and cotton in kharif season and wheat in Rabi season. The kharif crops account for 70 per cent of the cropped area.

The result of physical and chemical analysis of soil samples were given in Table No.XI. The major area of this district is occupied by deep black soils, which predominantly occur in the northern talukas of Akot, Belapur, Akola and Murtizapur. This type of soil is also found in the central portion of Washim and northwest portion of Mangrurpir. The light and coarse textured soils, locally known as 'Barad' or 'Murmad', occur in the hilly area of Murtizapur and Mangrurpir.

In all 155 algal taxa could be identified from Akola district. Of the total, 103 were of Cyanophyta, 20 belonged to Chlorophyta, 3 belonged to Xanthophyta, 5 belonged to Euglenophyta and 24 belonged to Bacillariophyta. The taxa with their species were *Microcystis* (1), *Chroococcus* (6), *Gloeocapsa* (1), *Aphanocapsa* (4) *Aphanothece* (5), *Synechococcus* (1), *Chlorogloea* (1), *Oscillatoria* (19), *Phormidium* (8), *Lyngbya* (10), *Schizothrix* (2), *Cylindrospermum* (2), *Nostoc* (10), *Anabaena* (7), *Aulosira* (1), *Plectonema* (2), *Camptylonemopsis* (1), *Scytonematopsis* (1), *Scytonema* (2), *Tolypothrix* (3), *Microchaete* (4), *Calothrix* (7), *Hapalosiphon* (2) *Westiellopsis* (1), *Fischerella* (1), *Stigonema* (1), *Chlamydomonas* (3), *Chlorococcum* (1), *Chlorella* (1), *Trochiacia* (1), *Ankistrodesmus* (1), *Scenedesmus* (3), *Ulothrix*

(1), *Microspora* (1), *Protococcus* (1), *Closterium* (2), *Cosmarium* (5), *Vaucheria* (3), *Euglena* (1), *Trachelomonas* (4), *Cyclotella* (1), *Fragillaria* (1), *Synedra* (1), *Achnanthes* (1), *Stauronsis* (1), *Navicula* (6), *Pinnularia* (2), *Amphora* (1), *Cymbella* (5), *Nitzschia* (4), *Surirella* (1).

The algal taxa which occurred with the high frequency (above 50%) in this district were *Chroococcus macrococcus, C. spelaeus, C. turgidus, Aphanocapsa biformis. A.fonticola, Oscillatoria amoena, O. curviceps, O. decolorata, O. princeps, O. subbrevis f. crass, O. salina, Phormidium foveolarum, P. jenkelianum, Lyngbya aerugineo-coerulea, L.lagerheimii, Schizothrin, tenuis, Cylindrospermum musicola v. longispora, Nostoc commune, N. calcicola, N. microscoicum, N.paludosum, Anabaena anomala, A. laxa, A.naviculoides, A. torulosa, A. variabilis, A. variabilis v. kashiensis, , Camptylonemopsis iyengarii, Scytonematopsis woronichinii, Scytonema fremyii, Tolypothrix bouteil1ei, T. byssoidea, Microchaetae ca1othrichoides, M. tenera, Calothrix brevissima v. moniliforms, C. epiphytica, C. membranacea, C. marchica v. intermedia, Hapalosiphon intricatus, H.welwitschii, Stigonema bormoides, Chlamydomonas globosa, Ch1orococcum humicolo, Scenedesmus, dimorphus, Ulothrix variabilis, Protococcus viridis, Closterium acerosum, C. parvulum, Cosmarium granatum, Fragillaria brevistriata f. elongata, Synedra affinis, Achnanthes delicatula, Navicula clavata, N . grivillei, Pinnularia microstauron, Cymbella austriaca, C. cymbiformis, Nitzschia dissipata, N.gracillia.*

The survey from the rice fields of Vidarbha reveals that quantitatively and qualitatively the Cyanophycean form dominated other groups of algae present. Out of the 8 districts, Nagpur district is found to be much rich in algae as compared to other districts, since from Nagpur district 211 algal taxa could be identified. On comparing the algal flora of paddy fields, it is observed that majority of the algae in the rice fields are common. There is small variation in the subterranean algal flora even- though more or less the climatic factor remains the same. Author accounts these differences to regional variability due to difference in soil characteristics such as soil pH, soil texture, mineral elements and organic matter present in the soil. To some extent it may also be due to the use of different fertilizers and pesticides used and remotely to irrigation sources and cultivation practices.

From the results of the soil analysis it is clear that the paddy field soils of Vidarbha region were alkaline. Hence naturally, the Cyanophycean forms dominated over other groups in all the districts. The occurrence of some fresh water forms in some fields, were due to the difference in the habitat in different seasons.

Out of the 27 taxa reported by Singh (1933) from the studies of 15 soil samples, *Chroococcus minutes*, *Phormidium foveolarum*, *Anabaena variabilis* and *Nostoc commune* were found in this investigation. Banerji (1935) and De (1939) made cultural studies on the rice field soils of Faridpur (now in East Pakistan). Banerji reported 9 species of Myxophyceae which were dominant in those soils. De reported the presence of *Phormidium foveolarum*, *Anabaena anomala*, *A. naviculoides*, *A. variabilis V. ellipsospora*, *A. torulosa* and *A. oryzae* in the paddy field which were also recorded in the present investigation. Singh (1939), while studying the algae in cultures of the rice fields of four districts of Uttar Pradesh, viz., Mirzapur, Varanasi, Gorakhpur and Basti, recorded 17 species of Chlorophyceae, 14 species of Myxophyceae and 12 species of diatoms. Out of the Myxophyceae recorded by him, *Chroococcus minutus*, *Oscillatoria proboscidea*, O. *terebriformis*,

Microcoleus chthonoplastes and *Cylindrospermum musicola* had also been obtained in the present study.

Gonzalves and Gangla (1949) studied the algal flora of five paddy fields of Bombay suburbs and obtained 51 species of algae consisting of 33 Myxophyceae, 10 Bacillariophyceae and 8 Chlorophyceae. Out of the blue-green algae recorded by them, sixteen species, viz. *Oscillatoria animalis, O. schultzii, Phormidium foveolarum, Lyngbya scotti, L. aerugineo-coerulea, Schizothrix tenuis, Schizothrix arenaria, Nostoc entophytum, N. linckia, N.humifusum, N. calcicola, Scytonema hofmanni, Microchaete calothrixoides, Hapalosiphon welwitschii* and *H. intricatus* were recorded from the paddy soils of Vidarbha also. Some of these species had been reported by other workers from different places (Mitra, 1951; Datta and Venkataraman, 1958; Marathe, 1963; Marathe and Navalkar, 1963 and Pandey, 1965). Mitra (1951) found a predominance of blue-green algae in the desiccated soils of paddy fields. Forty out of a total of 71 algal species reported by him belonged to Myxophyceae. Gupta (1966) reported the presence of *Nostoc paludosum* and *Calothrix membranacea* from the paddy soils of Allahabad. All these species could also be observed from the paddy fields of Vidarbha. Singh (1961) investigated the rice fields of Uttar Pradesh and Bihar in search of the dominant blue-green algae and recorded about 25 species of this group. According to him, in the rice fields, there is a predominant, widespread and universal growth of communities of blue green algae. The species reported by Singh were also noticed in the paddy fields of Vidarbha.

Marathe (1963, 1964, 1966) made valuable contributions on the soil algae of Bombay division. While studying the effect of fertilizers on the subterranean algal flora of paddy fields, she obtained as many as 30 species of Myxophyceae in the cultures.

Later, she (1966) studied the algal flora of 24 soil samples taken from cultivated fields. She obtained 122 species of algae, out of which 81 belonged to Myxophyceae. The present observation of algae of Vidarbha paddy fields also stands with the result obtained by her. Pandey (1965), while working extensively on the paddy fields of Uttar Pradesh, observed a preponderance of the Myxophycean algae over the others. He obtained a total of 74 species, both in cultures and from natural collections. Majority of the algae reported by him were found in the present investigation also.

The dominance of blue-green algae in the paddy fields of Vidarbha stands with the opinion of Subba Raju (1972) who reported that pH was the main influencing factor in the distribution of the green and blue-green algae and Chlorophyceae favoured acidic pH, while blue-green favaoured alkaline pH. Laloraya and Mitra (1974) made a survey of the blue-green algal flora of the paddy fields of east and south India. A total of 122 species was recorded by them from cultures and collections from natural conditions. The algal flora in the present investigation is similar, for most of its part, with that of the paddy fields studied by them.

Venkataraman and his research group (1975) at I.A.R.I. New Delhi, surveyed extensively the blue green algal flora of the paddy field soils in 15 States of India. The distribution of soils harbouring nitrogen fixing blue-green algae varied from, as low as 4 percent to as high as 80 percent (Aiyer, 1965; Materassi and Balloni, 1965; Watanabe, 1959; Singh, 1961 and Pande, 1965). Their survey showed that out of 2213 soil samples from rice fields, only about 33 percent were found to harbour nitrogen fixing forms. Prasad. (1978), while studying the pH tolerance of soil blue green algae of sixteen types of crop fields, obtained 216 taxa from cultures and natural collections.

Most of the species encountered in their investigation were capable of tolerating under range of pH. The genera belonging to Chroococcales reported by them were also noticed in this investigation. The presence of more Cyanophycean members with nitrogen fixing forms in majority of the fields stands with the above observation.

Author collected and identified the algal forms in different seasons and at different stages of paddy crop by culturing them in the laboratory.

Table II:

Paddy field algal flora of Vidarbha region

CYANOPHYTA

1. *Microcystis elabens* (Breb.) Kutz.
2. *Microcystis holsatica* lemmermann.
3. *Microcystis insignis* (Bech) Comb. Nov.
4. *Microcystis marginata* (Menegh.) Kutz.
5. *Microcystis protocystis* Crow
6. *Microcystis pulvera Var. incerta* (Lemm.) Crow.
7. *Microcystis rubusta* (Clark) Nygaard.
8. *Microcystis stagnalis* Lemm.
9. *Microcystis scripta* (Richter) Lemm.
10. *Chroococcus granulosus* Zeller.
11. *Chroococcus limneticus* Lemm.
12. *Chroococcus micrococcus* (Kutz.) Rabenh
13. *Chroococcus minutes* (Kutz) Nag.
14. *Chroococcus montanus* Hansgirg.

15. *Chroococcus minor* (Kutz) Nag.
16. *Chroococcus pallidus* Nag.
17. *Chroococcus schizodermaticus* West.
18. *Chroococcus spelaeus* Ercegovic.
19. *Chroococcus turgidus* (Kutz.) Nag.
20. *Chroococcus turgidus* (Kutz.) Nag. Var. *fuscescens* (Kutz.)
21. *Chroococcus tenax* (krichn.) Hieron.
22. *Chroococcus westii* (W. West) Boye-Petersen.
23. *Gloeocapsa atrata* (Turp.) Kutz.
24. *Gloeocapsa decorticans* (A. Br.) Richter.
24. *Gloeocapsa decorticans* (A. Br.) Richter.
25. *Gloeocapsa montana* Kutz.
26. *Gloeocapsa nigrescens* Nag.
27. *Gloeocapsa polydermatica* kutz.
28. *Gloeocapsa punctata* Nag.
29. *Gloeocapsa pleurocapsoides* Novacek
30. *Gloeocapsa rupestris* kutz.
31. *Glococapsa stegophila* (Itzigs.) Rebenh.
32. *Gloeothece membranacea* (Rabenh.) Bornet.
33. *Gloeothece palea* (Kutz.) Rebenh.
34. *Gloeothece rupestris* (Lyngb.) Barnet.
35. *Gloeothece samoensis* Wille.
36. *Gloeothece samoensis* Wille Var. Major Wille.
37. *Aphanocapsa biformis* A. Br.
38. *Aphanocapsa banaresensis* Bharadwaja.
39. *Aphanocapsa delicatissima* W. et. G.S. West.

40. *Aphanocapsa elachista* W. et. G.S. West.
41. *Aphanocapsa fonticola* Hansgirg.
42. *Aphanocapsa grevillei* (Hass.) Rabenh.
43. *Aphanocapsa montana* Cramer.
44. *Aphanocapsa muscicola* (Menegh.) Wille.
45. *Aphanocapsa nivalis* Lagerheim
46. *Aphanocapsa* pulchra (Kutz.) Rabenh.
47. *Aphanocapsa* roeseana de Bary
48. *Aphanothece clathrata* W. et. G.S. West.
49. *Aphanothece conferta* Richter.
50. *Aphanothece castagnei* (Breb.) Rabenh.
51. *Aphanothece microscopica* Nag.
52. *Aphanothece naegeli* Wartm.
53. *Aphanothece nidulans* Richter, P.
54. *Aphanothece Pallida* (Kutz.) Rabenh.
55. *Aphanothece stagnina* (Spreng.) A.Br.
56. *Aphanothece saxicola* Nag.
57. *Synechococcus aeruginosus* Nag.
58. *Synechococcus cedrorum* Sauvageau
59. *Synechococcus selngatus* Nag.
60. *Synechocystis aquatilis* Sauv.
61. *Synechocystis pevalekii* Ercegovic.
62. *Merismopedia elegans* A.Br.
63. *Merismopedia punctata* Meyen.
64. *Merismopedia tenuissima* Lemm.
65. *Chlorogloea microcystodies* Geitler.

66. *Chroococcidiopsis mysorensis* Tiwari, G.S
67. *Democarpa olivaea* (Reinsch) Tilden Var. *gigantean* Rao, C.B.
68. *Arthrospira balkrishnanii* Kamat.
69. *Arthrospira jenneri* Stizenb Ex. Gomont.
70. *Arthrospira khannae* Drouet et Strickland.
71. *Spirulina gigantean* Schmidle.
72. *Spirulina laxissima* West, G.S.
73. *Spirulina major* Kutz. Ex Gomont.
74. *Spirulina Subtilissima* Kutz. Ex Gomont.
75. *Oscillatoria anguina* (Bory) Gomont.
76. *Oscillatoria amoena* (Kutz.) Gomont.
77. *Oscillatoria animalis* Ag. Ex Gomont.
78. *Oscillatoria acuta* Bruhl et Biswasm orth. Mut. Geitler.
79. *Oscillatoria annae* van Goor.
80. *Oscillatoria amphibian* Ag. Ex Gomont.
81. *Oscillatoria amphigranulata* van Goor.
82. *Oscillatoria brevis* (Kutz.) Gomont.
83. *Oscillatoria biswasii* Kamat.
84. *Oscillatoria bharadwajae* Kamat.
85. *Oscillatoria curviceps* Ag. ex. Gomont.
86. *Oscillatoria curviceps* Var. *angusta* Ghose.
87. *Oscillatoria carakiana* playfair. V. non-granulata Kamat.
88. *Oscillatoria cortiana* Meneghini ex. Gomont.
89. *Oscillatoria chalybea* (mertens) Gomont var. *insularis* Gardner.

90. *Oscillatoria chalybea* (Mertens) Gomont V. *Minor* Kamat.
91. *Oscillatoria chilkensis* Biswas.
92. *Oscillatoria chlorine* kutz. ex Gomont.
93. *Oscillatoria decolorata* West, G.S.
94. *Oscillatoria foreaui* Fremy.
95. *Oscillatoria geitleriana* Elenkin.
96. *Oscillatoria grunowiana* Gomont.
97. *Oscillatoria irriqua* (Kutz.) Gomont.
98. *Oscillatoria jasorvensis* Vovk.
99. *Oscillatoria jasorvensis* Vovk.
100. *Oscillatoria limosa* Ag. ex. Gomont.
101. *Oscillatoria limosa* Ag. ex. Gomont. Var. *disperse-granulata* Schkorb.
102. *Oscillatoria limnetica* Lemm.
103. *Oscillatoria laete-virens* (Crouan) Gomont.
104. *Oscillatoria laete-virens* (Crouan) Gomont Var. *minimus* Biswas.
105. *Oscillatoria mougeotii* Kutz.
106. *Oscillatoria nigroviridis* Thwaites ex. Gomont.
107. *Oscillatoria nigra* Vaucher.
108. *Oscillatoria obscura* Bruhl et Biswas.
109. *Oscillatoria okeni* Ag. ex. Gomont.
110. *Oscillatoria ornata* Kutz. ex. Gomont.
111. *Oscillatoria princeps* Vaucher ex. Gomont.
112. *Oscillatoria proboscidea* Gomont.

113. *Oscillatoria profunda* Krrchner.
114. *Oscillatoria prolifica* (Grev.) Gomont.
115. *Oscillatoria pseudogeminata* G. Schmid.
116. *Oscillatoria quadripunctulata* Bruhl et. Biswas.
117. *Oscillatoria rubescens* DC ex. Gomont.
118. *Oscillatoria rubescens* DC ex. Gomont. Var. *Curvata* Kamat
119. *Oscillatoria raoi* De Toni, J.
120. *Oscillatoria raciborskii* Wolosz.
121. *Oscillatoria schultzii* Lemm.
122. *Oscillatoria schulzii* Lemm. V. *cylindrica* Kamat.
123. *Oscillatoria subuliformis* Kutz. ex. Gomont.
124. *Oscillatoria subbrevis* schmidle.
125. *Oscillatoria subbrevis* schmidle. f. *crassa* Dixit.
126. *Oscillatoria subbrevis* schmidle. f. *major* West, G.S.
127. *Oscillatoria subtilissima* Kutz.
128. *Oscillatoria splendida* Grev. ex. Gomont.
129. *Oscillatoria splendida* Grev. ex. Gomont. Var. *attenuata* W. and G.S. West.
130. *Oscillatoria salina* Biswas.
131. *Oscillatoria sancta* (Kutz.) Gomont.
132. *Oscillatoria terebriformis* Ag. ex. Gomont.
133. *Oscillatoria tenuis* Ag. ex. Gomont.
134. *Oscillatoria vizagapatensis* Rao, C.B.
135. *Phormidium africanum* Lemm.
136. *Phormidium angustissimum* W. et. G.S. West

137. *Phormidium anomala* Rao, C.B.
138. *Phormidium* amiguum Gomont.
139. *Phormidium abronema* Skuja.
140. *Phormidium bohneri* Schmidle.
141. *Phormidium ceylanicum* Wille.
142. *Phormidium ceylanicum* Wille. Var. *minor* Kamat.
143. *Phormidium corium* (ag.) Gomont. Var. *capitatum* Gardner.
144. *Phormidium cebennense* Gomont.
145. *Phormidium dimorphum* Lemm.
146. *Phormidium foveolarum* (Mont.) Gomont.
147. *Phormidium fragile* (Meneghini) Gomont.
148. *Phormidium favosum* (Bory) Gomont.
149. *Phormidium hieronymusii* Lemmermann.
150. *Phormidium jenkelianum* Schmid, G.
151. *Phormidium luridum* (Kutz.) Gomont.
152. *Phormidium mucosum* Gardner.
153. *Phormidium mucosum* Gardner. Var. *arvense* Rao, C.B.
154. *Phormidium molle* (Kutz.) Gomont.
155. *Phormidium molle* (Kutz.) Gomont. F. *tenuior* W. and G.S. West.
156. *Phormidium mucicola* Hub. – Pestalozzi et Naumann.
157. *Phormidium orientale* West, G.S.
158. *Phormidium papyraceum* (Ag.) Gomont.
159. *Phormidium stagnina* Rao, C.B.
160. *Phormidium tenue* (Menegh.) Gomont.
161. *Phormidium uncinatum* (Ag.) Gomont.

162. *Lyngbya attenuata* Fritsch.
163. *Lyngbya aestuarii* Liebm. ex. Gomont. Var. *tenuis* Dixit.
164. *Lyngbya allorgei* Fremy
165. *Lyngbya aerugineo-coerulea* (Kutz.) Gomont.
166. *Lyngbya cryptovaginata* Schkorbatow.
167. *Lyngbya ceylanica* Wille.
168. *Lyngbya corticicola* Bruhi et Biswas.
169. *Lyngbya contorta* Lemm.
170. *Lyngbya dendrobia* Bruhl et. Biswas.
171. *Lyngbya dendrobia* Bruhl et. Biswas f. *lurida* Bruhl et. Biswas Gomont.
172. *Lyngbya dendrobia* Bruhl et. Biswas Var. *skujaii* nom. nov.
173. *Lyngbya digueti* Gomont.
174. *Lyngbya gardneri* (Setchell et Gardner) Geitler
175. *Lyngbya hieronmusii* Lemm.
176. *Lyngbya Kuetzingii* Schmidle.
177. *Lyngbya kuetzingiana* Kirchner.
178. *Lyngbya lagerheimii* (Mob.) Gomont.
179. *Lyngbya lachneri* (zimmermann) Geitler.
180. *Lyngbya limnetica* Lemmermann.
181. *Lyngbya lutea* (Ag) Gom.
182. *Lyngbya major* Menegh ex. Gomont.
183. *Lyngbya martensiana* Menegh. ex. Gomont.
184. *Lyngbya martensiana* Menegh. ex. Gomont. Var. *minor* Gardner.

185. *Lyngbya maharastrensis* Kamat.
186. *Lyngbya majuscula* Harvey ex. Gomont.
187. *Lyngbya nordgardhii* Wille.
188. *Lyngbya palmarum* (Martens) Bruhl et. Biswas.
189. *Lyngbya polysiphoniae* fremy.
190. *Lyngbya perelegans* Lemm.
191. *Lyngbya putealis* Mont. ex. Gomont.
192. *Lyngbya rivularianum* Gom.
193. *Lyngbya stagnina* Kutzing.
194. *Lyngbya scotti* Fritsch.
195. *Lyngbya subtilis* West, W. Var. *granulosa* Bruhl and Biswas
196. *Lyngbya schmidlei* V. *major* Kamat.
197. *Lyngbya semoplena* (C.Ag.) J. Ag. ex. Gomont.
198. *Lyngbya shackletoni* W. et. G.S. West.
199. *Lyngbya truncicola* Ghose
200. *Schizothrix arenaria* (Berk.) Gomont.
201. *Schizothrix lardacea* (Ces.) Gomont.
202. *Schizothrix lacustris* A. Br. ex. Gomont.
203. *Schizothrix tenuis* Woronich.
204. *Symploca cartilaginea* (Mont.) Gomont.
205. *Symploca elegans* Kutzing ex. Gomont.
206. *Symploca muralis* Kutzing ex. Gomont.
207. *Symploca muscorum* (Ag.) Gomont.
208. *Symploca parietina* (A. Br.) Gomont.
209. *Microcoleus acutissimus* Gardner.

210. *Microcoleus amplus* Gardner
211. *Microcoleus chthonoplastes* Thurst ex. Gomont.
212. *Microcoleus paludosus* (Kutz.) Gomont.
213. *Hydrocoleus subincrustaceus.* Hansgrig.
214. *Anabaenopsis arnolodii* Aptekarj Var. *indica* Ramanathan.
215. *Anabaenopsis circularis* (G.S. West) Wolosz. et. Miller
216. *Anabaenopsis tanganyikae* (West, G.S.) Wolosz et. Miller
217. *Cylindrospermum gorakhpurense* Singh, R.N.
218. *Cylindrospermum indicum* Rao, C.B., orth. Mut. De. Toni.
219. *Cylindrospermum licheniforme* Kutz ex. Born. et. Flah.
220. *Cylindrospermum majus* Kutzing ex. Borm et. Flah.
221. *Cylindrospermum musicola* Kutzing ex. Borm et. Flah.
222. *Cylindrospermum musicola* Kutzing ex. Borm et. Flah.
223. *Cylindrospermum musicola* Kutzing ex. Borm et. Flah.
224. *Cylindrospermum michailovskoense* Elenkin.
225. *Nostoc coeruleum* Lyngbye ex Born. et. Flah.
226. *Nostoc commune* Vaucher ex Born. et. Flah.
227. *Nostoc calcicola* Brebisson ex Born. et. Flah.
228. *Nostoc ellipsosporm* (Desm.) Rabenh. ex Born. et. Flah.
229. *Nostoc entophytum* Born. et. Flah.
230. *Nostoc humifusum* Carmichael ex Born. et. Flah.
231. *Nostoc linckia* (Roth) Bornet ex Born. et. Flah.
232. *Nostoc microscopicum* Carm. ex Born. et. Flah.
233. *Nostoc muscorum* Ag. ex Born. et. Flah.
234. *Nostoc maculiforme* Born. et. Flah.

235. *Nostoc paludosum* Kutzing ex Born. et. Flah.
236. *Nostoc piscinale* Kutzing ex Born. et. Flah.
237. *Nostoc punctiforme* (Kutz.) Harjot.
238. *Nostoc spongiaeforme* Agardh ex Born. et. Flah.
239. *Nostoc sphaericum* Vaucher ex Born. et. Flah.
240. *Anabaena anomala* Fritsch.
241. *Anabaena ambigua* Rao., C.B.
242. *Anabaena circunalis* Rabenhorst ex Born. et. Flah.
243. *Anabaena doliolum* Bharadwaja.
244. *Anabaena fertilissima* Rao, C.B.
245. *Anabaena Iyengarii* Bharadwaja.
246. *Anabaena Iyengarii* Bharadwaja Var. *tenuis* Rao, C.B.
247. *Anabaena laxa* (Rabehn.)
248. *Anabaena naviculoides* Fritsch.
249. *Anabaena oryzae* Fritsch.
250. *Anabaena oscillariodes* Bory ex Born. et. Flah.
251. *Anabaena oscillariodes* Bory ex Born. et. Flah. Var. *angustus* Bharadwaja.
252. *Anabaena orientalis* Dixit.
253. *Anabaena randhawe* Venkataraman, G.S.
254. *Anabaena sphaerica* Bornet et. Flahault.
255. *Anabaena sphaerica* Bornet et. Flahault. Var. *attenuata* Bharadwaja.
256. *Anabaena torulosa* (Carm.) Lagerh. ex Born. et. Flah.
257. *Anabaena vaginicola* Fristsch et Rich.
258. *Anabaena variabilis* Kutzing ex Born. et. Flah.

259. *Anabaena variabilis* Var. *ellipsospora.*

260. *Anabaena variabilis* Var. *kashiensis.*

261. *Pseudanabaena catenata* Lauterb.

262. *Pseudanabaena schmidlei* Jaag. O.

263. *Pseudanabaena schmidlei* Jaag. O. f. *gracilis* Skuja.

264. *Nodularia spumigena* mertens ex Born. et. Flah.

265. *Aulosira aenigmatica* Fremy.

266. *Aulosira fertilissima* Ghose.

267. *Aulosira fertilissima* Ghose Var. *tenuis* Rao, C.B.

268. *Aulosira* implexa Bornet et. Flah.

269. *Aulosira prolific* Bharadwaja.

270. *Aulosira pseudoramosa* Bharadwaja.

271. *Plectonema gracillimum* (Zopf) Hansgirg.

272. *Plectonema hansgirgi* schmidle.

273. *Plectonema nostocorum* Bornet. ex Gomont.

274. *Plectonema puteale* (Kirchner) Hansgirg.

275. *Plectonema tomasinianum* (Kutz) Born. et. Gomant.

276. *Camptylonemopsis iyengarii* Desikachary.

277. *Scytonematopsis woronichinni* Kiss.

278. *Scytonema cincinnatum* Thuret ex. Born. et Flah.

279. *Scytonema fremyii* non. nov.

280. *Scytonema hofmanni* Ag. ex. Born. et Flah.

281. *Scytonema javanicum* (Kutz.) Bornet ex. Born. et Flah.

282. *Scytonema millei* Bornet ex. Born. et Flah.

283. *Scytonema ocellatum* Lungbye ex. Born. et Flah.

284. *Scytonema pseudohofmanni* Bharadwaja.

285. *Scytonema schmidlei* J. De. Toni.
286. *Tolypothrix bouteillei* (Berb. Et. Desm.) Forti.
287. *Tolypothrix byssoidea* (Berk.) Kirchner.
288. *Tolypothrix nodosa* Bharadw.
289. *Microchaete aegualis* (Fremy) Comb. nov.
290. *Microchaete calothrichoides* Hansgirg.
291. *Microchaete loktakensis* Bruhi et Biswas.
292. *Microchaete tenera* Thuret ex. Born. et Flah.
293. *Microchaete tenera* Thuret ex. Born. et Flah.
294. *Microchaete uberrima* Carter, N.
295. *Microchaete uberrima* Carter, N. f. *minor* Carter.
296. *Fortiea incerta* Skuja.
297. *Calothrix braunii* (A.Br.) Bornet. et Flahault.
298. *Calothrix brevissima* West, G.S.
299. *Calothrix brevissima* West, G.S. var. *moniliforme* Ghose orth. Mut. Geitler.
300. *Calothrix castellii* (Massal.) Born. et Flah.
301. *Calothrix clavata* West, G.S.
302. *Calothrix* epiphytica W. et West, G.S.
303. *Calothrix elenkinii* Kossinskaja.
304. *Calothrix javanica* de Wilde.
305. *Calothrix membranacea* Schmidle.
306. *Calothrix marchica* Lemmermann.
307. *Calothrix marchica* Lemmermann. Var. *crassa* Rao, C.B.
308. *Calothrix marchica* Lemmermann. Var. *intermedia* Rao, C.B.

309. *Dichothrix baueriana* (Grun.) Born. et Flah.
310. *Rivularia beccariana* (De not.) Born. et Flah.
311. *Rivularia bullata* (Poir) Berk. ex Born. et Flah.
312. *Gloeothrichia indica* Schmidle.
313. *Gloeothrichia intermedia* (Lemm.) Geitler Var. *kanwaensis* Rao, C.B.
314. *Gloeothrichia natans* Rabenhorst ex Born. et Flah.
315. *Nostochopsis lobatus* Wood em. Geitler.
316. *Hapalosiphon baronii* W. et. G.S. West.
317. *Hapalosiphon hansgirgi* (Schmidle) Forti.
318. *Hapalosiphon intricatus* W. et. G.S. West.
319. *Hapalosiphon welwitschii* W. et. G.S. West.
320. *Westiellopsis prolifica* Janet.
321. *Fischerella muscicola* (Thuret) Gom.
322. *Stigonema hormoides* (Kutz.) Born, et. Flah.
323. *Stigonema lavardei* Fremy Var. *indica* Fremy.

CHLOROPHYTA :

1. *Chlamydomonas globosa* Snow.
2. *Chlamydomonas lismorensis* Playton.
3. *Chlamydomonas mucicola* Schmidle.
4. *Chlamydomonas reinhardi* Dangeard
5. *Chlamydomonas sphaginicola*. Fritsch et. Takeda.
6. *Carteria klebsii* (Dang.) Dill.
7. *Characium ambiguum*. Hermann ex Rabnchhorst.
8. *Chlorococcum humicolo* (Naegeli) Rabnehhorst.

9. *Chlorococcum infusionum* (Schrank) Meneghini
10. *Chlorococcum vitiosum.* Printz.
11. *Trebouxia humicolo* (Treboux) West. Et. Fritsch.
12. *Chlorella conglomerata* (Artari) oltamanns.
13. *Chlorella vulgaris* Beijerinck.
14. *Tetraedron quadratum* (Reinsch) Hansgirg. F. *minus.* (Reinsch) De Toni.
15. Tetraedron muticum (A. Braun.) Hansgirg.
16. *Tetraedron octaedricum* v. *spinosum* (Reinsch.) W. et. G.S. West.
17. *Trochiscia aciculifera* (Lagerb.) Hansgirg.
18. *Trochiscia granulata* (Reinsch.) Hansgirg.
19. *Trochiscia reticularis* (Reinsch.) Hansgirg.
20. *Oocystis elliptica* W. West.
21. *Oocystis lacustris.* Chodat.
22. *Oocystis nodulosa* West et West.
23. *Protosiphon botryoides* (Kutz.) Keibs.
24. Ankistrodesmus falcatus (Corda) Ralfs.
25. *Ankistrodesmus falcatus* (Corda) Ralfs. V. *tumidus.* (West et West) G.S. West.
26. *Selenastrum gracile* Reinsch.
27. *Scenedesmus arcuatus* (Lemmermann) Limmermann.
28. *Scenedesmus bijugatus* (Turpin) Kutzing Var. *bicellularis* (Chodat) comb. nov.
29. *Scenedesmus bijugatus* (Turpin) Kutzing Var. *irregularis* (Wille) Smith.
30. *Scenedesmus* dimorphus. (Turpin) Kutzing.

31. Scenedesmus *falcatus* Chodat.
32. *Scenedesmus longus* v. breviseina. Smith.
33. *Scenedesmus oblicuus.* (Turpin) Kuetzing.
34. *Scenedesmus quadricaude* v. maximus W. et. G.S. West.
35. *Ulothrix oscillarina* Kuetzing.
36. *Ulothrix subtilissima* Rabenhorst.
37. *Ulothrix tenissima* Kuetzing.
38. *Ulothrix tenerrima* (Kuetzing) Kuetzing.
39. *Ulothrix variabilis* (Kuetzing) Kuetzing.
40. *Ulothrix zonata* (Weber et Mohr) Kuetzing.
41. *Geminella minor* (Naegeli) Heering.
42. *Geminella protogenita* (Kuetzing) G.S. West.
43. *Microspora abbreviate* (Rabenhorst) Lagerheim.
44. *Microspora floccosa* (Vaucher) Thuret.
45. *Microspora stagnorum* (Kutzing) Lagerheim.
46. *Rhizoclonium heiroglyphicum.* (Agardh) Kuetzing.
47. *Rhizoclonium hookeri* Kuetzing.
48. *Protococcus viridis* Ag. (Prescott) Smith. F.
49. *Chaetophora elegans* (Roth) Ag.
50. *Chaetophora pisiformis* Ag.
51. *Chaetophora scutata* de Brebisson
52. *Bulbochaete brevifulta* Wittrock ex Hirn.
53. *Oedocladium indicum* Kamat.
54. *Oedogonium aquaticum* Kamat.
55. *Oedogonium formosum* Kamat.
56. *Oedogonium globoum* Nordstedt ex. Hirn

57. *Oedogonium longipilum* Jao.
58. *Oedogonium unisporum* Kamat.
59. *Oedogonium maharastrensis* Kamat.
60. *Oedogonium princeps* (Hassall) Wittrock ex. Hirn.
61. *Oedogonium rugulosum* Nordstodt ex Hirn.
62. *Spirogyra arta* Jao.
63. *Spirogyra biformis* Jao.
64. *Spirogyra communis* (Hassal) Kutzing.
65. *Spirogyra chakiense* Kolkwitz and Krieger.
66. *Spirogyra dictyospora* Jao.
67. *Spirogyra ghosei* Singh.
68. *Spirogyra hyaline* Cleve.
69. *Spirogyra irregularis* Nageli.
70. *Spirogyra inflata* (Vaucher) Kuetzing.
71. *Spirogyra maxima* (Hassall) Wittrock.
72. *Spirogyra prolifica* kamat.
73. *Spirogyra submaxima* Transeau.
74. *Spirogyra turfosa* Gay.
75. *Sirocladium maharashtranse* Randhawa.
76. *Sirogonium ventersicum* Transeau.
77. *Closterium acerosum* (Schrank) Ehrenberg.
78. *Closterium acutum* (Lyngbye) de Brebisson.
79. *Closterium langeolatum* Kuetzing.
80. *Closterium paryulum* Nag.
81. *Closterium venus* Kuetzing.
82. *Cosmarium contractum* Kirchner.

83. *Cosmarium* depressum (Naeg.) Lund. Var. *Achondrum* (Boldt) West et. West.
84. *Cosmarium dentiforum* Corda Var. *minus* Kamat.
85. *Cosmarium granatum* Brebisson.
86. *Cosmarium impressulum* Elfv.
87. *Cosmarium meneghinii* de Brebisson.
88. *Cosmarium subimpressulum*. Borge.
89. *Cosmarium trilobatum* Reinsch.
90. *Cosmarium undulatum* Corda ex Ralfs.

XANTHOPHYTA :

1. *Vaucheria amphibian* Randhawa.
2. *Vaucheria geminąta* (Vauch.) De Condolle.
3. *Vaucheria hamata* Walz.
4. *Vaucheria mayyanadensis* Erady.
5. *Vaucheria sessilis* (Vauch) De Condolle.
6. *Botrydium* species.

EUGLENOPHYTA :

1. *Euglena* acus Ehrenberg v. *hyalina* Klebs.
2. *Euglena charkowiensis* Swir.
3. *Euglena gracillis* Klebs.
4. *Euglena multiformis* Schiller.
5. *Euglena tuberculata* Swirenko.
6. *Euglena fusca* (Klebs) Lemm.
7. *Euglena maharastrensis* Kamat.
8. *Phacus acuminatus* stokes v. *discifera* (Pochm) Huber-Pestalozzi.

9. *Phacus caudatus* Huebner.
10. *Phacus hamatus* Pochmann.
11. *Phacus longicauda* (E) Dujardin.
12. *Phacus* platalea Drez. v. minor kamath.
13. Phacus *horridus* Pochmann.
14. *Phacus curvicauda* Swirenko.
15. *Phacus tortus* (Lemm.) Skvorkzov.
16. *Trachelomonas allia* Drez. em. Deflandre.
17. *Trachelomonas* hispida (Perty) Stein emnd. Deflandre.
18. *Trachelomonas lefevrei* Deflandre.
19. *Trachelomonas oblonga* Lemmermann.
20. *Trachelomonas volvocina* Ehrenberg.
21. *Trachelomonas volvocina* Ehrenberg v. *derephora* Conrad.
22. *Trachelomonas armata* (E) stein v. *steinii* emend. Deflandre.
23. *Trachelomonas cordata* (Drez.) comb. Deflandre f. *minor* Deflandre.
24. *Trachelomonas hispida* (Perty) Stein emend. Deflandre f. *minor* Bourr.
25. *Trachelomonas pulcherrima* Playfair.

BACILLARIOPHYTA :

1. *Cyclotella kutzingiana* Thwaites.
2. *Cyclotella operculata* Kg.
3. *Fragillaria brevistriata* f. elongata Venk.
4. *Synedra affinis*. Kg.
5. *Synedra pulchella* var. *lanceolata* o. Meara.
6. *Synedra ulna* (Nitzsch.) Ehrenb.

7. *Achnanthes coarctata* Berb.
8. *Achnanthes delicatula* Kg.
9. *Achnanthes microcephala.* Kg.
10. *Gyrosigma spenceri* (W.Sm.) Cl.
11. *Stauroneis anceps* Ehrenh.
12. *Navicular anceps* (Ehrenb.) var. *linearis* Ehrenb. (als Art).
13. *Navicula aponina* Kg.
14. *Navicula bacilliformis* Grun.
15. *Navicula brebissonii* Kg.
16. *Navicula clavata* Greg.
17. *Navicula crucicula* (W. Sm.) Donk.
18. *Navicula cryptocephala* Kg. f. *terrestris* Lund.
19. *Navicula gregaria* Dunk.
20. *Navicula grivillei* (Ag.) Cl.
21. *Navicula lanceolata* (Ag.) Kg.
22. *Navicula major* Kg.
23. *Navicula mutica* Kutz.
24. *Navicula sculpta* Ehernb.
25. *Navicula salinarum* Grun.
26. *Navicula terrestris* Petersen.
27. *Navicula tennelloides* (Hust.) emend.
28. *Navicula vaucheriae* var. *densistriata* Lund.
29. *Pinnularia microstauron* (Ehr.) Cleve.
30. *Pinnularia viridis* (Nitzsch.) Ehr.
31. *Amphora caffaeiformis* Ag.
32. *Amphora ovalis* Kg.

33. *Amphora montana* (Krasake).
34. *Cymbella alpine* Grun.
35. *Cymbella austriaca* Grun.
36. *Cymbella cymbiformis* (Kg.) V.H.
37. *Cymbella gracilis* (Rebenh.) Cleve.
38. *Cymbella parva* W. Sm
39. *Cymbella reinhardtii* Grun.
40. *Cymbella tumescens.* A. Cl.
41. *Cymbella ventricosa* (kg.) Cleve.
42. *Rhopalodia gibba* (Ehrenb.) O. Muller Var. *Ventricosa* (Enrenb.) Grun.
43. *Gomphonema angustatum* (Kg.) Grun. var. *producta* Grun.
44. *Gomphonema gracile* Ehernb.
45. *Gomphonema lanceolatum* Ehr.
46. *Hantzschia anphioxys* Grun.
47. *Nitzschia angustata* (W.Sm.) Grun..
48. *Nitzschia dissipata* (Kg.) Grun.
49. *Nitzschia gracilis* Hantzsch.
50. *Nitzschia hantzschiana* Rabenh.
51. *Nitzschia intermedia* Hantzsch.
52. *Nitzschia kuetzingiana* Hilse.
53. *Nitzschia palea* (Kg.) W.Sm.
54. *Nitzschia parvula* Lewis var. *terricola* Lund.
55. *Nitzschia vermicularis* (Kg.) Hantzsch.
56. *Surirella baltica* Schum.

Table III:

Algal Flora of Vidarbha Paddy fields with their State of occurrence.

Places of Occurrence

Sr. No	*Name of Algae observed*	*Algae observed in soil cultures*	*Algae observed in water bodies*	*Algae observed in wet soil surface*
1	2	3	4	5
	CYANOPHYTA			
1	*Microcystis elabens*	P	-	-
2	*Microcystis holsatica*	P	P	-
3	*Microcystis insignis*	-	P	-
4	*Microcystis marginata*	-	P	-
5	*Microcystis protocystis*	-	P	-
6	*Microcystis pulvera Var. incerta*	P	-	P
7	*Microcystis rubusta*	P	-	-
8	*Microcystis stagnalis*	-	P	-
9	*Microcystis scripta*	P	-	-
10	*Chroococcus granulosus*	P	-	P
11	*Chroococcus limneticus*	P	-	-
12	*Chroococcus micrococcus*	P	-	-
13	*Chroococcus minutes*	-	-	P
14	*Chroococcus montanus*	P	-	-
15	*Chroococcus minor*	P	-	-

Contd...

Sr. No	Name of Algae observed	Algae observed in soil cultures	Algae observed in water bodies	Algae observed in wet soil surface
16	*Chroococcus pallidus*	-	-	P
17	*Chroococcus schizodermaticus*	-	P	-
18	*Chroococcus spelaeus*	P	-	-
19	*Chroococcus turgidus*	P	-	-
20	*Chroococcus turgidus var. fuscescens*	P	-	-
21	*Chroococcus tenax*	-	P	-
22	*Chroococcus westii*	-	-	P
23	*Gloeocapsa atrata*	-	-	P
24	*Gloeocapsa decorticans*	P	-	-
25	*Gloeocapsa Montana*	P	-	-
26	*Gloeocapsa nigrescens*	P	-	-
27	*Gloeocapsa polydermatica*	P	-	P
28	*Gloeocapsa punctata*	P	-	-
29	*Gloeocapsa pleurocapsoides*	P	-	-
30	*Gloeocapsa rupestris*	-	-	P
31	*Gloeocapsa stegophila*	P	-	-
32	*Gloeothece membranacea*	P	-	-
33	*Gloeothece palea*	-	P	-
34	*Gloeothece rupestris*	-	-	P
35	*Gloeothece samoensis*	P	-	-

Contd...

Sr. No	Name of Algae observed	Algae observed in soil cultures	Algae observed in water bodies	Algae observed in wet soil surface
36	*Gloeothece samoensis var. major*	P	-	-
37	*Aphanocapsa biformis*	P	-	P
38	*Aphanocapsa banaresensis*	P	-	-
39	*Aphanocapsa delicatissima*	-	P	-
40	*Aphanocapsa elachista*	-	P	-
41	*Aphanocapsa fonticola*	P	-	-
42	*Aphanocapsa grevillei*	-	P	-
43	*Aphanocapsa Montana*	P	-	-
44	*Aphanocapsa muscicola*	-	P	-
45	*Aphanocapsa nivalis*	P	-	-
46	*Aphanocapsa pulchra*	-	-	P
47	*Aphanocapsa roeseana*	P	-	-
48	*Aphanothece clathrata*	-	P	-
49	*Aphanothece conferta*	P	-	P
50	*Aphanothece castagnei*	-	P	-
51	*Aphanothece microscopic*	P	-	-
52	*Aphanothece naegeli*	P	-	-
53	*Aphanothece nidulans*	P	-	-
54	*Aphanothece pallida*	-	-	P
55	*Aphanothece stagnina*	P	-	-
56	*Aphanothece saxicola*	P	-	-
57	*Synechococcus aeruginosus*	-	-	P

Contd...

Sr. No	Name of Algae observed	Algae observed in soil cultures	Algae observed in water bodies	Algae observed in wet soil surface
58	*Synechococcus cedrorum*	P	-	-
59	*Synechococcus selngatus*	P	P	-
60	*Synechocystis aquatilis*	-	P	-
61	*Synechocystis pevalekii*	-	-	P
62	*Merismopedia elegans*	-	P	-
63	*Merismopedia punctata*	P	-	-
64	*Merismopedia tenuissima*	P	-	-
65	*Chlorogloea microcystodies*	P	-	-
66	*Chroococcidiopsis mysorensis*	P	-	-
67	*Democarpa olivaea var. gigantean*	P	-	-
68	*Arthrospira balkrishnanii*	P	-	-
69	*Arthrospira jenneri*	-	P	-
70	*Arthrospira khannae*	P	-	-
71	*Spirulina gigantean*	-	P	-
72	*Spirulina laxissima*	P	-	-
73	*Spirulina major*	P	-	-
74	*Spirulina subtilissima*	-	-	P
75	*Oscillatoria anguina*	-	-	P
76	*Oscillatoria amoena*	P	-	-
77	*Oscillatoria animalis*	-	-	P
78	*Oscillatoria acuta*	P	-	-

Contd...

Sr. No	Name of Algae observed	Algae observed in soil cultures	Algae observed in water bodies	Algae observed in wet soil surface
79	*Oscillatoria annae*	-	-	P
80	*Oscillatoria amphibian*	-	P	-
81	*Oscillatoria amphigranulata*	-	-	P
82	*Oscillatoria brevis*	P	-	-
83	*Oscillatoria biswasii*	-	-	P
84	*Oscillatoria bharadwajae*	P	-	-
85	*Oscillatoria curviceps*	P	-	-
86	*Oscillatoria curviceps var. angusta*	P	-	-
87	*Oscillatoria carakiana v. non-granulata*	P	-	-
88	*Oscillatoria cortiana*	-	-	P
89	*Oscillatoria chalybea var. insularis*	P	-	-
90	*Oscillatoria chalybea var. minor*	P	-	-
91	*Oscillatoria chilkensis*	P	-	-
92	*Oscillatoria chlorine*	-	-	P
93	*Oscillatoria decolorata*	P	-	-
94	*Oscillatoria foreaui*	P	-	-
95	*Oscillatoria geitleriana*	P	-	-
96	*Oscillatoria grunowiana*	P	-	-

Contd...

Sr. No	*Name of Algae observed*	*Algae observed in soil cultures*	*Algae observed in water bodies*	*Algae observed in wet soil surface*
97	*Oscillatoria irriqua*	P	-	-
98	*Oscillatoria jasorvensis*	P	-	-
99	*Oscillatoria jenensis*	P	-	-
100	*Oscillatoria limosa*	-	-	P
101	*Oscillatoria limosa var. disperso-granulata*	P	-	-
102	*Oscillatoria limnetica*	P	-	-
103	*Oscillatoria laete-virens*	-	-	-
104	*Oscillatoria laete-virens var. minimus*	P	-	-
105	*Oscillatoria mougeotii*	P	-	-
106	*Oscillatoria nigroviridis*	P	-	-
107	*Oscillatoria nigra*	P	-	-
108	*Oscillatoria obscura*	P	-	-
109	*Oscillatoria okeni*	-	-	P
110	*Oscillatoria ornate*	-	-	P
111	*Oscillatoria princeps*	P	-	P
112	*Oscillatoria proboscidea*	P	-	-
113	*Oscillatoria profunda*	P	-	-
114	*Oscillatoria prolific*	P	-	-
115	*Oscillatoria pseudogeminata*	P	-	-
116	*Oscillatoria quadripunctulata*	P	-	P

Contd...

Sr. No	Name of Algae observed	Algae observed in soil cultures	Algae observed in water bodies	Algae observed in wet soil surface
117	*Oscillatoria rubescens*	-	-	P
118	*Oscillatoria rubescens var. curvata*	P	-	-
119	*Oscillatoria raoi*	P	-	-
120	*Oscillatoria raciborskii*	P	-	-
121	*Oscillatoria schultzii*	P	-	-
122	*Oscillatoria schulzii var. cylindrical*	P	-	-
123	*Oscillatoria subuliformis*	P	-	-
124	*Oscillatoria subbrevis*	-	-	P
125	*Oscillatoria subbrevis f. crassa*	P	-	-
126	*Oscillatoria subbrevis f. major*	P	-	-
127	*Oscillatoria subtilissima*	P	-	-
128	*Oscillatoria splendid*	P	-	-
129	*Oscillatoria splendida var. attenuate*	P	-	-
130	*Oscillatoria salina*	P	-	-
131	*Oscillatoria sancta*	P	-	-
132	*Oscillatoria terebriformis*	P	-	-
133	*Oscillatoria tenuis*	P	-	-
134	*Oscillatoria vizagapatensis*	P	-	-
135	*Phormidium africanum*	P	-	-

Contd...

Sr. No	Name of Algae observed	Algae observed in soil cultures	Algae observed in water bodies	Algae observed in wet soil surface
136	*Phormidium angustissimum*	-	P	-
137	*Phormidium anomala*	P	-	-
138	*Phormidium amiguum*	-	P	-
139	*Phormidium abronema*	-	-	P
140	*Phormidium bohneri*	-	-	P
141	*Phormidium ceylanicum*	P	-	-
142	*Phormidium ceylanicum var. minor*	P	-	-
143	*Phormidium corium var. capitatum*	P	-	-
144	*Phormidium cebennense*	P	-	-
145	*Phormidium dimorphum*	-	-	P
146	*Phormidium foveolarum*	P	-	P
147	*Phormidium fragile*	-	-	P
148	*Phormidium favosum*	P	-	-
149	*Phormidium hieronymusii*	P	-	-
150	*Phormidium jenkelianum*	P	-	P
151	*Phormidium luridum*	P	-	-
152	*Phormidium mucosum*	P	-	-
153	*Phormidium mucosum var. arvense*	P	-	-
154	*Phormidium molle*	-	P	-

Contd...

Sr. No	Name of Algae observed	Algae observed in soil cultures	Algae observed in water bodies	Algae observed in wet soil surface
155	*Phormidium molle f. tenuior*	P	-	-
156	*Phormidium mucicola*	-	P	-
157	*Phormidium orientale*	P	-	-
158	*Phormidium papyraceum*	P	-	-
159	*Phormidium stagnina*	-	P	-
160	*Phormidium tenue*	P	-	-
161	*Phormidium uncinatum*	P	-	-
162	*Lyngbya attenuata*	P	-	-
163	*Lyngbya aestuarii var. tenuis*	P	-	-
164	*Lyngbya allorgei*	P	-	-
165	*Lyngbya aerugineo-coerulea*	P	P	-
166	*Lyngbya cryptovaginata*	-	P	-
167	*Lyngbya ceylanica*	-	-	P
168	*Lyngbya corticicola*	-	P	P
169	*Lyngbya contorta*	P	-	-
170	*Lyngbya dendrobia*	P	-	-
171	*Lyngbya dendrobia f. lurida*	P	-	-
172	*Lyngbya dendrobia var. skujaii*	P	-	-
173	*Lyngbya digueti*	-	-	P

Contd...

Sr. No	Name of Algae observed	Algae observed in soil cultures	Algae observed in water bodies	Algae observed in wet soil surface
174	*Lyngbya gardneri*	P	-	-
175	*Lyngbya hieronmusii*	-	P	-
176	*Lyngbya Kuetzingii*	P	-	-
177	*Lyngbya kuetzingiana*	P	-	-
178	*Lyngbya lagerheimii*	-	P	-
179	*Lyngbya lachneri*	P .	-	-
180	*Lyngbya limnetica*	P	-	-
181	*Lyngbya lutea*	P	-	-
182	*Lyngbya major*	P	-	-
183	*Lyngbya martensiana*	P	-	-
184	*Lyngbya martensiana var. minor*	P	-	-
185	*Lyngbya maharastrensis*	P	-	-
186	*Lyngbya majuscula*	P	-	-
187	*Lyngbya nordgardhii*	-	P	-
188	*Lyngbya palmarum*	P	-	-
189	*Lyngbya polysiphoniae*	P	-	-
190	*Lyngbya perelegans*	-	P	-
191	*Lyngbya putealis*	-	P	-
192	*Lyngbya rivularianum*	-	-	P
193	*Lyngbya stagnina*	P	-	-
194	*Lyngbya scotti*	P	-	-
195	*Lyngbya subtilis var. granulose*	P	-	-

Contd...

Sr. No	Name of Algae observed	Algae observed in soil cultures	Algae observed in water bodies	Algae observed in wet soil surface
196	*Lyngbya schmidlei var. major*	P	-	-
197	*Lyngbya semiplena*	P	-	-
198	*Lyngbya shackletoni*	P	-	-
199	*Lyngbya truncicola*	P	-	-
200	*Schizothrix arenaria*	P	-	P
201	*Schizothrix lardacea*	P	-	-
202	*Schizothrix lacustris*	-	P	-
203	*Schizothrix tenuis*	P	-	P
204	*Symploeca cartilaginea*	-	-	P
205	*Symploca elegans*	P	-	-
206	*Symploca muralis*	-	-	P
207	*Symploca muscorum*	P	-	-
208	*Symploca parietina*	-	-	P
209	*Microcoleus acutissimus*	P	-	-
210	*Microcoleus amplus*	P	-	-
211	*Microcoleus chthonoplastes*	P	-	-
212	*Microcoleus paludosus*	-	-	P
213	*Hydrocoleus subincrustaceus.*	P	-	-
214	*Anabaenopsis arnoldii*	P	-	-
215	*Anabaenopsis circularis*	P	-	-
216	*Anabaenopsis tanganyikae*	-	P	-

Contd...

Sr. No	Name of Algae observed	Algae observed in soil cultures	Algae observed in water bodies	Algae observed in wet soil surface
217	*Cylindrospermum gorakhpurense*	-	-	P
218	*Cylindrospermum indicum*	-	-	P
219	*Cylindrospermum licheniforme*	P	-	P
220	*Cylindrospermum majus*	-	P	P
221	*Cylindrospermum musicola*	P	-	P
222	*Cylindrospermum musicola var. longispora*	P	P	-
223	*Cylindrospermum musicola var. kashmiriensis*	-	P	-
224	*Cylindrospermum michailovskoense*	-	-	P
225	*Nostoc coeruleum*	P	P	-
226	*Nostoc commune*	P	-	P
227	*Nostoc calcicola*	P	-	P
228	*Nostoc ellipsosporum*	-	P	-
229	*Nostoc entophytum*	P	-	P
230	*Nostoc humifusum*	-	P	-
231	*Nostoc linckia*	P	-	P
232	*Nostoc microscopicum*	P	-	P
233	*Nostoc muscorum*	-	-	P
234	*Nostoc maculiforme*	P	-	-
235	*Nostoc paludosum*	P	P	-

Contd...

Sr. No	**Name of Algae observed**	**Algae observed in soil cultures**	**Algae observed in water bodies**	**Algae observed in wet soil surface**
236	*Nostoc piscinale*	-	-	P
237	*Nostoc punctiforme*	-	P	-
238	*Nostoc spongiaeforme*	-	P	-
239	*Nostoc sphaericum*	P	-	-
240	*Anabaena anomala*	P	-	P
241	*Anabaena ambigua*	P	-	-
242	*Anabaena circunalis var. crassa*	P	-	-
243	*Anabaena doliolum*	-	P	-
244	*Anabaena fertilissima*	P	-	-
245	*Anabaena iyengarii*	-	P	-
246	*Anabaena iyengarii var. tenuis*	P	-	P
247	*Anabaena laxa*	P	P	-
248	*Anabaena naviculoides*	P	-	P
249	*Anabaena oryzae*	P	-	P
250	*Anabaena oscillariodes*	P	P	-
251	*Anabaena oscillariodes var. angustus*	-	P	-
252	*Anabaena orientalis*	P	-	-
253	*Anabaena randhawe*	P	-	-
254	*Anabaena sphaerica*	P	P	-
255	*Anabaena sphaerica var. attenuate*	-		P
256	*Anabaena torulosa*	P	P	-

Contd...

Sr. No	Name of Algae observed	Algae observed in soil cultures	Algae observed in water bodies	Algae observed in wet soil surface
257	*Anabaena vaginicola*	P	-	-
258	*Anabaena variabilis*	P	-	P
259	*Anabaena variabilis var. ellipsospora*	P	-	-
260	*Anabaena variabilis var. kashiensis*	P	P	-
261	*Pseudanabaena catenata*	-	P	-
262	*Pseudanabaena schmidlei*	P	-	-
263	*Pseudanabaena schmidlei f. gracilis*	P	-	-
264	*Nodularia spumigena var. major*	-	P	-
265	*Aulosira aenigm[illegible]a*	-	P	-
266	*Aulosira fertilissima*	-	P	-
267	*Aulosira fertilissima var. tenuis*	P	-	-
268	*Aulosira implexa*	P	-	-
269	*Aulosira prolifica*	P	P	-
270	*Aulosira pseudoramosa*	-	P	-
271	*Plectonema gracillimum*	-	P	-
272	*Plectonema hansgirgi*	-	-	P
273	*Plectonema nostocorum*	P	-	P
274	*Plectonema puteale*	P	-	-
275	*Plectonema tomasinianum*	-	P	-

Contd...

Sr. No	Name of Algae observed	Algae observed in soil cultures	Algae observed in water bodies	Algae observed in wet soil surface
276	*Camptylonemopsis iyengarii*	P	-	P
277	*Scytonematopsis woronichinni*	P	P	-
278	*Scytonema cincinnatum*	P	-	-
279	*Scytonema fremyii*	P	-	P
280	*Scytonema hofmanni*	-	-	P
281	*Scytonema javanicum*	P	-	P
282	*Scytonema millei*	P	-	-
283	*Scytonema ocellatum var. puropureum*	P	-	-
284	*Scytonema pseudohofmanni*	-	-	P
285	*Scytonema schmidlei*	P	-	-
286	*Tolypothrix bouteillei*	P	-	P
287	*Tolypothrix byssoidea*	P	-	P
288	*Tolypothrix nodosa*	-	P	-
289	*Microchaete aequalis*	P	P	-
290	*Microchaete calothrichoides*	P	P	-
291	*Microchaete loktakensis*	-	P	-
292	*Microchaete tenera*	P	-	P
293	*Microchaete tenera Var. tenuis*	-	P	-

Contd...

Sr. No	Name of Algae observed	Algae observed in soil cultures	Algae observed in water bodies	Algae observed in wet soil surface
294	*Microchaete uberrima*	-	P	-
295	*Microchaete uberrima f. minor*	P	-	-
296	*Fortiea incerta*	-	P	-
297	*Calothrix braunii*	P	-	P
298	*Calothrix brevissima*	-	P	-
299	*Calothrix brevissima var. moniliforme*	P	P	-
300	*Calothrix castellii*	P	-	-
301	*Calothrix clavata*	-	P	-
302	*Calothrix epiphytica*	P	P	-
303	*Calothrix elenkinii*	-	P	-
304	*Calothrix javanica*	-	-	P
305	*Calothrix membranacea*	P	-	P
306	*Calothrix marchica*	P	-	-
307	*Calothrix marchica var. crassa*	-	P	-
308	*Calothrix marchica var. intermedia*	P	P	-
309	*Dichothrix baueriana*	P	-	-
310	*Rivularia beccariana*	P	-	P
311	*Rivularia beccariana*	P	-	P
312	*Gloeothrichia indica*	-	P	-

Contd...

Sr. No	Name of Algae observed	Algae observed in soil cultures	Algae observed in water bodies	Algae observed in wet soil surface
313	*Gloeothrichia intermedia var kanwaensis*	-	P	-
314	*Gloeothrichia natans*	P	-	-
315	*Nostochopsis lobatus*	-	P	-
316	*Hapalosiphon baronii*	-	P	-
317	*Hapalosiphon hansgirgi*	-	P	-
318	*Hapalosiphon intricatus*	P	P	-
319	*Hapalosiphon welwitschii*	P	-	P
320	*Westiellopsis prolifica*	P	-	P
321	*Fischerella muscicola*	P	-	-
322	*Stigonema hormoides*	P	-	P
323	*Stigonema lavardei var. indica*	-	-	P
	CHLOROPHYTA			
1	*Chlamydomonas globosa*	P	P	-
2	*Chlamydomonas lismorensis*	-	P	-
3	*Chlamydomonas mucicola*	P	-	-
4	*Chlamydomonas reinhardi*	P	-	-
5	Chlamydomonas *sphaginicola.*	-	P	-
6	*Carteria klebsii*	P	-	-
7	*Characium ambiguum.*	-	-	P

Contd...

Sr. No	*Name of Algae observed*	*Algae observed in soil cultures*	*Algae observed in water bodies*	*Algae observed in wet soil surface*
8	*Chlorococcum humicolo*	P	-	P
9	*Chlorococcum infusionum*	P	-	P
10	*Chlorococcum vitiosum.*	-	P	-
11	*Trebouxia humicolo*	-	P	-
12	*Chlorella conglomerata*	P	-	-
13	*Chlorella vulgaris*	P	-	P
14	*Tetraedron guadratum* f. *minus*	P	-	-
15	*Tetraedron muticum*	-	P	-
16	Tetraedron *octaedricum* var. *spinosum*	-	P	-
17	*Trochiscia aciculifera*	-	P	-
18	*Trochiscia granulate*	P	-	-
19	*Trochiscia reticularis*	-	P	-
20	*Oocystis elliptica*	-	P	-
21	*Oocystis lacustris.*	-	P	-
22	*Oocystis* nodulosa	-	P	-
23	*Protosiphon botryoides*	-	P	-
24	*Ankistrodesmus falcatus*	-	P	-
25	*Ankistrodesmus falcatus* var. *tumidus*	-	-	P
26	*Selenastrum gracile*	-	P	-
27	*Scenedesmus arcuatus*	P	-	-

Contd...

Sr. No	Name of Algae observed	Algae observed in soil cultures	Algae observed in water bodies	Algae observed in wet soil surface
28	*Scenedesmus bijugatus* var. *bicellularis*	-	P	-
29	*Scenedesmus bijugatus* var. *irregularis*	-	-	P
30	*Scenedesmus dimorphus.*	P	-	P
31	*Scenedesmus falcatus*	-	P	-
32	*Scenedesmus longus* var. *brevispina*	-	P	-
33	*Scenedesmus oblicuus.*	-	P	-
34	*Scenedesmus quadricaude* var. *maximus*	-	P	-
35	*Ulothrix oscillarina*	-	P	-
36	*Ulothrix subtilissima*	-	P	-
37	*Ulothrix* tenissima	-	P	-
38	*Ulothrix tenerrima*	P	-	-
39	*Ulothrix variabilis*	P	-	P
40	*Ulothrix zonata*	-	P	-
41	*Geminella minor*	P	-	-
42	*Geminella protogenita*	P	-	-
43	*Microspora abbreviate*	-	P	-
44	*Microspora floccosa*	-	P	-
45	*Microspora stagnorum*	-	P	-
46	*Rhizoclonium heiroglyphicum.*	-	P	-

Contd...

Sr. No	Name of Algae observed	Algae observed in soil cultures	Algae observed in water bodies	Algae observed in wet soil surface
47	*Rhizoclonium hookeri*	-	P	-
48	*Rotococcus viridis*	P	-	-
49	*Chaetophora elegans*	-	P	-
50	*Chaetophora pisiformis*	-	P	-
51	*Coleochaete acutata*	-	P	-
52	*Bulbochaete brevifulta*	-	P	-
53	*Oedocladium indicum*	-	P	-
54	*Oedogonium aquaticum*	-	P	-
55	*Oedogonium formosum*	-	P	-
56	*Oedogonium globoum*	-	P	-
57	*Oedogonium longipilum*	-	P	-
58	*Oedogonium unisporum*	-	P	-
59	*Oedogonium maharastrensis*	-	P	-
60	*Oedogonium princeps*	-	P	-
61	*Oedogonium rugulosum*	-	P	-
62	*Spirogyra arta*	-	P	-
63	*Spirogyra biformis*	-	P	-
64	*Spirogyra communis*	-	P	-
65	*Spirogyra chakiense*	-	P	-
66	*Spirogyra dictyospora*	-	P	-
67	*Spirogyra ghosei*	-	P	-
68	*Spirogyra hyaline*	-	P	-

Contd...

Sr. No	Name of Algae observed	Algae observed in soil cultures	Algae observed in water bodies	Algae observed in wet soil surface
69	*Spirogyra irregularis*	-	P	-
70	*Spirogyra inflate*	-	P	-
71	*Spirogyra maxima*	-	P	-
72	*Spirogyra prolific*	-	P	-
73	*Spirogyra submaxima*	-	P	-
74	*Spirogyra turfosa*	-	P	-
75	*Sirocladium maharashtranse*	-	P	-
76	*Sirogonium ventersicum*	-	P	-
77	*Closterium acerosum*	P	P	-
78	*Closterium acutum*	-	P	-
79	*Closterium langeolatum*	-	P	-
80	*Closterium parvulum*	P	-	P
81	*Closterium venus*	-	-	P
82	*Cosmarium contractum*	-	P	-
83	*Cosmarium depressum var. achondrum*	-	P	-
84	*Cosmarium dentiforum var. minus*	-	-	P
85	*Cosmarium granatum*	P	-	P
86	*Cosmarium impressulum*	-	-	P
87	*Cosmarium meneghinii*	-	P	-
88	*Cosmarium subimpressulum.*	-	P	-

Contd...

Sr. No	*Name of Algae observed*	*Algae observed in soil cultures*	*Algae observed in water bodies*	*Algae observed in wet soil surface*
89	*Cosmarium trilobatum*	-	P	-
90	*Cosmarium undulatum*	-	-	P
	XANTHOPHYTA :			
1	*Vaucheria amphibian*	-	-	P
2	*Vaucheria geminate*	-	-	P
3	*Vaucheria hamata*	-	-	P
4	*Vaucheria mayyanadensis*	-	-	P
5	*Vaucheria sessilis*	-	-	P
6	*Botrydium species.*	P	-	-
	EGLENOPHYTA			
1	*Euglena acus* var. *hyaline*	-	P	-
2	*Euglena charkowiensis*	-	P	-
3	*Euglena gracillis*	-	P	-
4	*Euglena multiformis*	-	P	-
5	*Euglena tuberculata*	-	P	-
6	*Euglena fusca*	-	P	-
7	*Euglena maharastrensis*	-	P	-
8	*Phacus acuminatus* var. *discifera*	-	P	-
9	*Phacus caudatus*	-	P	-
10	*Phacus hamatus*	-	P	-
11	*Phacus longicauda*	-	P	-
12	*Phacus platalea* var. *minor*	-	P	-

Contd...

Sr. No	Name of Algae observed	Algae observed in soil cultures	Algae observed in water bodies	Algae observed in wet soil surface
13	*Phacus horridus*	-	P	-
14	*Phacus* curvicauda	-	P	-
15	*Phacus tortus*	-	P	-
16	*Trachelomonas allia*	-	P	-
17	*Trachelomonas hispida*	-	P	-
18	*Trachelomonas lefevrei*	-	P	-
19	*Trachelomonas oblonga*	-	P	-
20	*Trachelomonas volvocina*	-	P	-
21	*Trachelomonas volvocina* var. *derephora*	-	P	-
22	*Trachelomonas armata* var. *steinii*	-	P	-
23	*Trachelomonas cordata* f. *minor*	-	P	-
24	*Trachelomonas hispida* f. *minor*	-	P	-
25	*Trachelomonas pulcherrima*	-	P	-
	BACILLARIOPHYTA			
1	*Cyclotella kutzingiana*	-	P	-
2	*Cyclotella operculata*	-	P	-
3	*Fragillaria brevistriata* f. *elongate*	P	P	-
4	*Synedra affinis*	P	P	-

Contd...

Sr. No	*Name of Algae observed*	*Algae observed in soil cultures*	*Algae observed in water bodies*	*Algae observed in wet soil surface*
5	*Synedra pulchella* var. *lanceolata*	P	P	-
6	*Synedra ulna*	P	P	-
7	*Achnanthes coarctata*	P	P	-
8	*Achnanthes delicatula*	P	P	-
9	*Achnanthes microcephala*	P	P	-
10	*Gyrosigma spenceri*	P	P	-
11	*Stauroneis anceps*	P	P	-
12	*Navicular anceps* var. *linearis*	P	P	-
13	*Navicula aponina*	P	P	-
14	*Navicula bacilliformis*	P	P	-
15	*Navicula brebissonii*	P	P	-
16	*Navicula clavata*	P	P	-
17	*Navicula crucicula*	P	P	-
18	*Navicula cryptocephala* f. *terrestris*	P	P	-
19	*Navicula gregaria*	P	P	-
20	*Navicula grivillei*	P	P	-
21	*Navicula lanceolata*	P	P	-
22	*Navicula major*	P	P	-
23	*Navicula mutica*	P	P	-
24	*Navicula sculpta*	P	P	-

Contd...

Sr. No	Name of Algae observed	Algae observed in soil cultures	Algae observed in water bodies	Algae observed in wet soil surface
25	*Navicula salinarum*	P	P	-
26	*Navicula terrestris*	P	P	-
27	*Navicula tennelloides*	P	P	-
28	*Navicula vaucheriae* var. *densistriata*	P	P	-
29	*Pinnularia microstauron*	P	P	-
30	*Pinnularia viridis*	P	P	-
31	*Amphora caffaeiformis*	P	P	-
32	*Amphora ovalis*	P	P	-
33	*Amphora montana*	P	P	-
34	*Cymbella alpina*	P	P	-
35	*Cymbella austriaca*	P	P	-
36	*Cymbella cymbiformis*	P	P	-
37	*Cymbella gracilis*	P	P	-
38	*Cymbella parva*	P	P	-
39	*Cymbella reinhardtii*	P	P	-
40	*Cymbella tumescens*	P	P	-
41	*Cymbella ventricosa*	P	P	-
42	*Rhopalodia gibba* var. *ventricosa*	P	P	-
43	*Gomphonema angustatumvar. producta*	P	P	-
44	*Gomphonema gracile*	P	P	-

Contd...

Sr. No	*Name of Algae observed*	*Algae observed in soil cultures*	*Algae observed in water bodies*	*Algae observed in wet soil surface*
45	*Gomphonema lanceolatum*	P	P	-
46	*Hantzschia anphioxys*	P	P	-
47	*Nitzschia angustata*	P	P	-
48	*Nitzschia dissipata*	P	P	-
49	*Nitzschia gracilis*	P	P	-
50	*Nitzschia hantzschiana*	P	P	-
51	*Nitzschia intermedia*	P	P	-
52	*Nitzschia kuetzingiana*	P	P	-
53	*Nitzschia palea*	P	P	-
54	*Nitzschia parvula* var. *terricola*	P	P	-
55	*Nitzschia vermicularis*	P	P	-
56	*Surirella baltica*	P	P	-

Table IV:

Result of Soil analysis
Name of District : Nagpur
Places of Sample

Constituents	*Nagpur*	*Kamptee*	*Ramtek*	*Katol*	*Kalmeshwar*	*Saoner*	*Kuhi*	*Umrer*	*Butibori*	*Narkhed*
	1	2	3	4	5	6	7	8	9	10
I. Physical analysis (Expressed as % on oven dry basis)										
Moisture content	6.0	11.20	6.0	12.0	11.0	9.5	10.5	10.7	10.5	10.0
Coarse sand	12.5	7.6	12.5	7.4	8.0	4.6	4.5	4.5	4.3	4.6
Fine sand	53.0	13.5	53.0	12.5	11.0	17.0	16.5	17.0	16.5	17.0
Silt	13.25	19.85	13.25	20.1	21.0	19.0	20.0	20.0	20.0	21.0
Clay	15.0	54.45	15.0	55.0	54.0	57.0	56.0	55.5	56.0	55.0
II. Chemical analysis										
Soil pH	8.0	7.9	8.0	7.3	7.6	7.2	7.8	6.8	7.8	7.2
Electric conductivity	0.22	0.25	0.22	0.3	0.4	0.25	0.30	0.15	0.30	0.24
Organic carbon	0.48	0.51	0.48	0.43	0.32	0.30	0.33	0.3	0.33	0.41
Total Nitrogen	0.048	0.051	0.048	0.043	0.032	0.03	0.033	0.03	0.033	0.041
K_2O kg/ha	193.0	264.0	193.0	202.0	198.0	180.0	204.0	170.0	190.0	200.0
P_2O_5 kg/ha	80.0	76.0	80.0	78.0	79.0	11.0	12.0	10.0	12.0	11.0
$CaCO_3$	5.0	7.0	5.0	6.0	5.3	7.0	8.0	7.0	8.0	7.0

Table V:

Result of Soil analysis
Name of District : Bhandara
Places of Sample

Constituents	*Bhandara*	*Tumsar*	*Tiroda*	*Gondia*	*Amgaon*	*Salekasa*	*Lakhani*	*Sakoli*	*Arjuni*	*Lakhandur*
	1	2	3	4	5	6	7	8	9	10
I. Physical analysis (Expressed as % on oven dry basis)										
Moisture content	3.0	6.0	6.0	5.0	8.0	5.0	9.0	5.0	10.0	7.0
Coarse sand	4.48	15.5	22.5	10.7	22.1	22.74	14.66	19.3	15.7	20.0
Fine sand	30.02	29.25	28.53	31.52	27.4	28.62	28.49	24.13	34.0	32.0
Silt	30.75	18.5	19.75	20.0	21.31	12.75	27.5	24.5	10.75	16.0
Clay.	31.75	30.5	24.5	33.25	22.5	30.75	20.0	25.75	30.0	28.75
II. Chemical analysis										
Soil pH	7.3	6.7	7.0	7.2	6.7	6.9	6.66	7.1	7.6	7.2
Electric conductivity	0.25	0.9	0.13	0.2	0.25	0.15	0.15	0.25	0.15	0.21
Organic carbon	0.35	0.49	0.25	0.4	0.24	0.25	0.50	0.60	0.75	0.68
Total Nitrogen	0.035	0.049	0.025	0.04	0.024	0.025	0.05	0.06	0.075	0.068
K_2O kg/ha	215.0	250.0	282.0	185.0	170.0	182.0	204.0	363.0	204.0	258.0
P_2O_5kg/ha	36.0	40.0	8.0	25.0	24.0	16.0	6.0	44.0	62.0	52.0
$CaCO_3$	2.0	2.0	4.0	5.0	4.0	5.0	2.0	1.5	1.0	3.0

Table VI:

Result of Soil analysis
Name of District : Chandrapur
Places of Sample

Constituents	*Sindewahi*	*Nagbhid*	*Bramhapuri*	*Armori*	*Chimur*	*Warora*	*Sironcha*	*Mool*	*Gadchiroli*
	1	2	3	4	5	6	7	8	9
I. Physical analysis (Expressed as % on oven dry basis)									
Moisture content	3.0	7.08	10.5	8.47	7.6	8.0	5.0	12.5	11.0
Coarse sand	4.9	10.6	4.97	12.32	19.0	21.2	17.6	15.5	14.6
Fine sand	23.0	20.0	21.78	10.77	23.1	29.1	31.77	24.7	26.0
Silt	29.0	30.25	32.0	38.5	17.2	15.24	16.5	18.0	19.0
Clay	40.0	35.75	37.25	33.0	29.0	28.0	30.5	29.0	28.0
II. Chemical analysis									
Soil pH	8.2	8.1	7.6	8.2	7.8	7.7	7.6	7.3	7.8
Electric conductivity	0.35	0.25	0.10	0.35	0.4	0.3	0.3	0.4	0.32
Organic carbon	0.32	0.27	0.55	0.24	0.43	0.53	0.22	0.46	0.52
Total Nitrogen	0.032	0.027	0.055	0.024	0.043	0.053	0.022	0.046	0.052
K_2O kg/ha	261.0	692.0	556.0	330.0	208.0	330.0	182.0	204.0	242.0
P_2O_5kg/ha	8.0	11.0	12.0	10.0	24.0	34.0	16.0	26.0	22.0
$CaCO_3$	5.0	4.0	6.5	5.0	2.8	1.5	2.5	2.7	1.9

Table VII:

Result of Soil analysis
Name of District : Wardha
Places of Sample

Constituents	*Wardha*	*Pulgaon*	*Hinganghat*	*Arvi*	*Karanja*
	1	2	3	4	5
I. Physical analysis (Expressed as % on oven dry basis)					
Moisture content	5.0	9.0	3.0	5.0	14.0
Coarse sand	10.02	4.12	8.99	8.78	12.2
Fine sand	28.05	14.43	31.44	12.11	15.9
Silt	30.0	40.0	24.25	28.25	12.25
Clay	28.25	30.75	33.75	45.1	46.5
II. Chemical analysis					
Soil pH	8.1	8.1	8.0	8.0	7.4
Electric conductivity	0.15	0.10	0.10	0.10	0.15
Organic carbon	0.52	0.45	0.15	0.48	0.30
Total Nitrogen	0.052	0.045	0.015	0.048	0.03
K_2O kg/ha	385.0	318.0	273.0	330.0	204.0
P_2O_5kg/ha	30.0	6.0	2.0	26.0	20.0
$CaCO_3$	9.0	3.5	11.0	5.0	3.0

Table VIII:

Result of Soil analysis
Name of District : Amravati
Places of Sample

Constituents	*Amravati*	*Badnera*	*Dhamangaon*	*Morshi*
	1	2	3	4
I. Physical analysis (Expressed as % on oven dry basis)				
Moisture content	8.0	3.0	4.0	3.0
Coarse sand	16.86	10.08	13.4	21.07
Fine sand	10.3	12.5	15.2	24.14
Silt	18.0	30.75	25.0	19.75
Clay	45.3	44.5	43.0	27.25
II. Chemical analysis				
Soil pH	8.0	7.8	8.1	7.9
Electric conductivity	0.21	0.38	0.15	0.7
Organic carbon	0.55	0.48	0.51	0.59
Total Nitrogen	0.055	0.048	0.051	0.059
K_2O kg/ha	390.0	238.0	295.0	261.0
P_2O_5 kg/ha	2.0	36.0	16.0	22.0
$CaCO_3$	7.0	10.5	8.0	8.0

Table IX:

Result of Soil analysis
Name of District : Yavatmal
Places of Sample

Constituents	*Yavatmal*	*Pusad*	*Kalamba*	*Wani*	*Khandala*
	1	2	3	4	5
I. Physical analysis (Expressed as % on oven dry basis)					
Moisture content	6.73	6.68	8.99	4.74	2.5
Coarse sand	10.3	5.2	3.8	3.14	15.3
Fine sand	20.98	18.95	25.9	28.05	34.0
Silt	24.0	30.0	20.5	23.5	20.0
Clay	40.0	43.0	46.25	44.0	30.5
II. Chemical analysis					
Soil pH	7.7	7.0	7.6	7.5	7.3
Electric conductivity	0.14	0.15	0.13	0.10	0.14
Organic carbon	0.41	0.33	0.22	0.15	0.69
Total Nitrogen	0.041	0.033	0.022	0.015	0.069
K_2O kg/ha	261.0	227.0	340.0	308.0	410.0
P_2O_5 kg/ha	56.0	46.0	54.0	14.0	48.0
$CaCO_3$	3.5	4.5	2.5	2.0	1.5

Table X:

Result of Soil analysis
Name of District : Buldana
Places of Sample

Constituents	*Khamgaon*	*Malkapur*	*Jalgaon jamod*	*Buldana*
	1	2	3	4
I. Physical analysis (Expressed as % on oven dry basis)				
Moisture content	4.0	11.0	8.32	3.0
Coarse sand	13.3	6.9	12.30	11.82
Fine sand	21.77	32.2	31.22	33.19
Silt	20.5	24.0	20.75	20.5
Clay	42.0	32.0	35.5	36.0
II. Chemical analysis				
Soil pH	8.1	8.2	7.6	8.1
Electric conductivity	0.28	0.22	0.6	0.15
Organic carbon	0.45	0.27	0.21	0.45
Total Nitrogen	0.045	0.027	0.021	0.045
K_2O kg/ha	204.0	421.0	398.0	427.0
P_2O_5 kg/ha	8.0	22.0	70.0	6.0
$CaCO_3$	4.5	3.5	1.5	3.5

Table XI:

Result of Soil analysis
Name of District : Akola
Places of Sample

Constituents	*Akola*	*Akot*	*Murtizapur*	*Malgaon*	*Umari*	*Pathur*
	1	2	3	4	5	6
I. Physical analysis (Expressed as % on oven dry basis)						
Moisture content	9.71	11.6	5.16	7.44	2.39	8.9
Coarse sand	5.5	14.48	8.25	10.3	10.5	10.1
Fine sand	30.0	26.37	35.0	22.9	23.6	20.5
Silt	23.75	14.75	22.0	20.0	30.0	35.75
Clay	40.00	35.75	30.4	45.0	35.0	31.0
II. Chemical analysis						
Soil pH	7.2	7.9	8.0	7.8	7.3	7.2
Electric conductivity	0.20	0.26	0.17	0.10	0.10	0.23
Organic carbon	0.18	0.48	0.33	0.18	0.18	0.41
Total Nitrogen	0.018	0.048	0.033	0.018	0.018	0.041
K_2O kg/ha	91.0	93.0	82.0	227.0	227.0	98.0
P_2O_5kg/ha	6.0	50.0	20.0	18.0	18.0	84.0
$CaCO_3$	6.5	8.0	2.0	5.0	5.0	6.0

PLATE – I

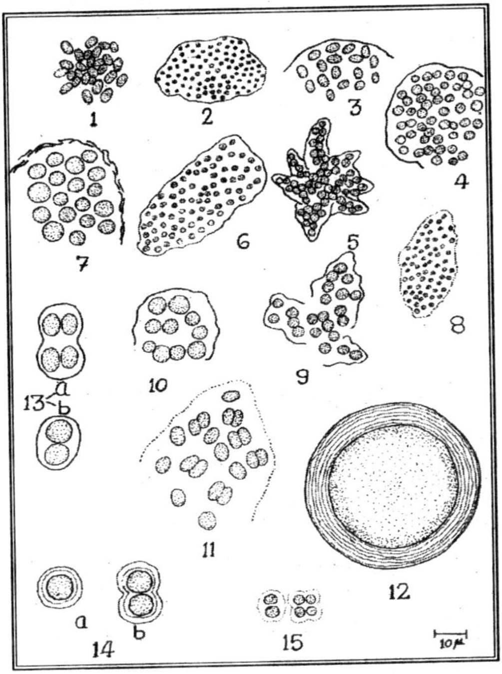

1. *Microcystis elabens* (Breb.) Kutz.
2. *Microcystis holsatica* Lemmermann.
3. *Microcystis insignis* (Bech.) Comb. nov.
4. *Microcystis marginata (menegh.) Kutz.*
5. *Microcystis protocystis* Crow.
6. *Microcystis pulvera* Var. incerta (Lemm.) Crow.
7. *Microcystis robusta* (Clark) Nygaard.
8. *Microcystis stagnalis* Lemm.
9. *Microcystis scripta* (Richter) Lemm.
10. *Chroococcus granulosus* Zeller.
11. *Chroococcus limneticus* Lemm.
12. *Chroococcus macrococcus* (Kutz.) Rabenh.
13. *Chroococcus minutus* (Kutz.) Nag.
14. *Chroococcus montanus* Hansgirg.
15. *Chroococcus minor* (Kutz.) Nag.

PLATE – II

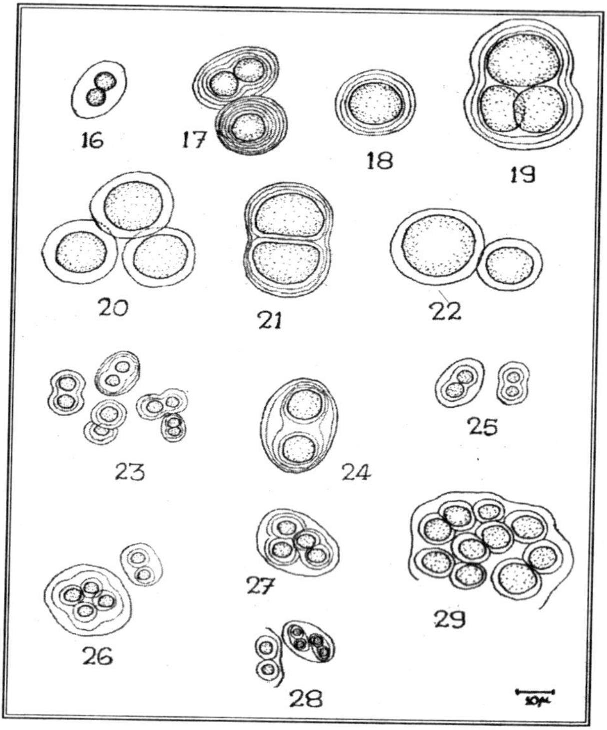

16. *Chroococcus pallidus* Nag.
17. *Chroococcus schizodermaticus* West.
18. *Chroococcus spelaeus* Ercegovic.
19. *Chroococcus turgidus* (Kutz.) Nag.
20. *Chroococcus turgidus* (Kutz.) Nag. Var. *fuscescens* (Kutz.).
21. *Chroococcus tenax* (Kirchn.) Hieron.
22. *Chroococcus westii* (W.West) Boye-Petersen.
23. *Gloeocapsa atrata* (Turp.) Kutz.
24. *Gloeocapsa decorticans* (A.Br.) Richter.
25. *Gloeocapsa montana* Kutz.
26. *Gloeocapsa nigrescens* Nag.
27. *Gloeocapsa polydermatica* Kutz.
28. *Gloeocapsa punctata* Nag.
29. *Gloeocapsa pleurocapsoides* Novacek.

PLATE – III

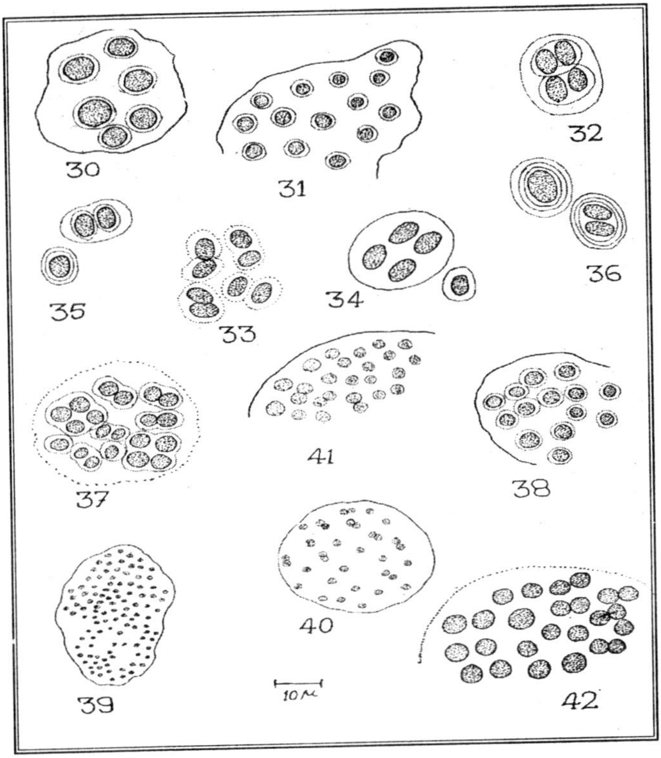

30. *Gloeocapsa rupestris* Kutz.
31. *Gloeocapsa stegophila* (Itzigs.) Rebenh.
32. *Gloeocapsa membranacea* (Rabenh.) Bornet.
33. *Gloeocapsa palea* (Kutz.) Rabenh.
34. *Gloeocapsa rupestris* (Lyngb.) Barnet.
35. *Gloeocapsa samoensis* Wille.
36. *Gloeocapsa samoensis* Wille var. *Major* Wille.
37. *Aphanocapsa biformis* A.Br.
38. *Aphanocapsa banaresensis* Bharadwaja.
39. *Aphanocapsa delicatissima* W. et. G. S. West.
40. *Aphanocapsa elachista* W. et. G.S. West.
41. *Aphanocapsa fonticola* Hansgirg.
42. *Aphanocapsa grevillei* (Hass.) Rabenh.

PLATE – IV

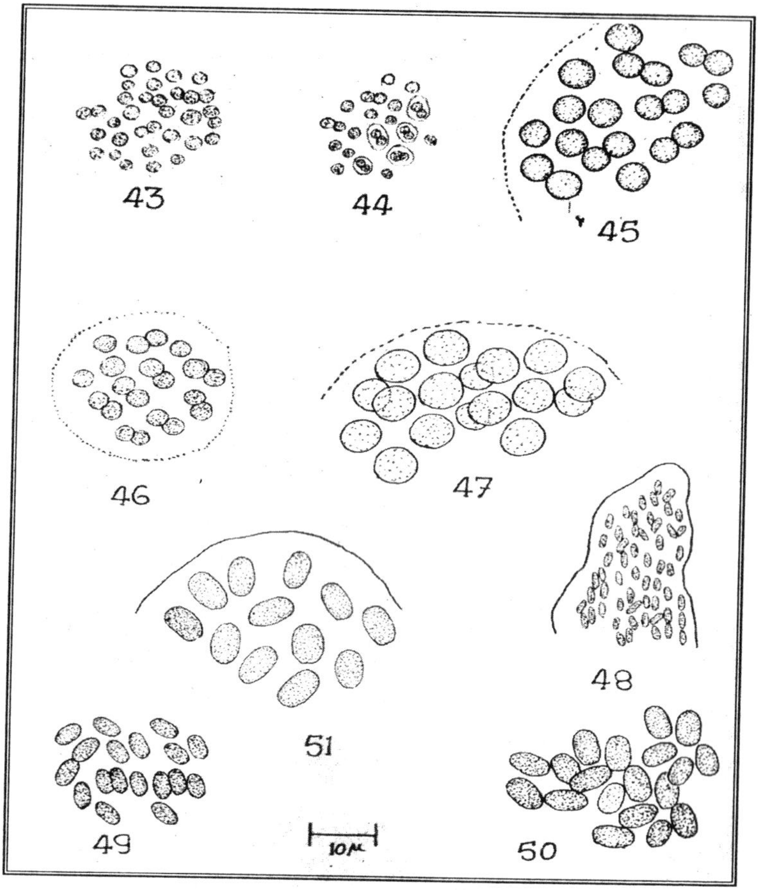

43. *Aphanocapsa montana* Cramer.
44. *Aphanocapsa muscicola* (Menegh.) Wille.
45. *Aphanocapsa nivalis* Lagerheim.
46. *Aphanocapsa pulchra* (Kutz.) Rabenh.
47. *Aphanocapsa roeseana* de Bary.
48. *Aphanothece clathrata.* W. et. G.8. West.
49. *Aphanothece conferta* Richter.
50. *Aphanothece castaqnei* (Breb.) Rabenh.
51. *Aphanothece microscopica* Nag.

PLATE – V

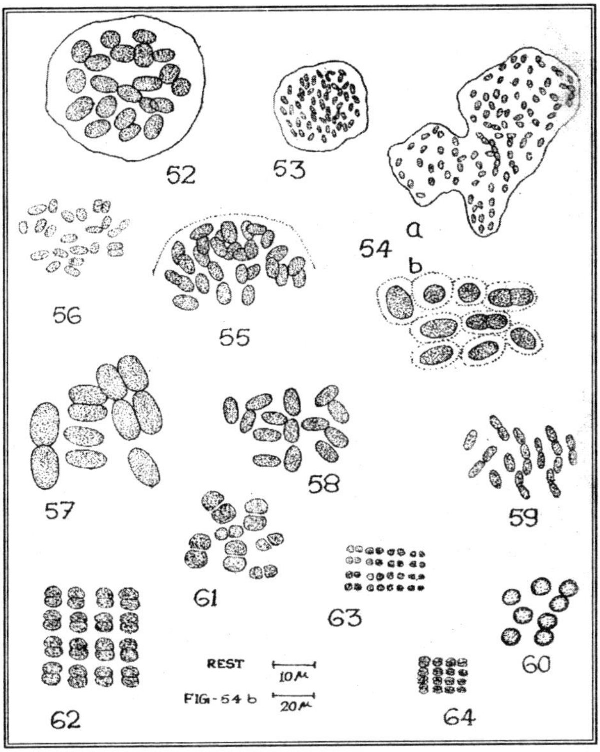

52. *Aphanothece naegelii* Wartm.
53. *Aphanothece nidulans* Richter, p.
54. *Aphanothece pallida* (Kutz.) Rabenh.
55. *Aphanothece stagnina* (Spreng.) A.Br.
56. *Aphanothece saxicola* Nag.
57. *Synechococcus aeruginosus* Nag.
58. *Synechococcus cedrorum* Sauvageau.
59. *Synechococcus elongatus* Nag.
60. *Synechococcus aquatilis* Sauv.
61. *Synechococcus pevalekii* Ercegovic.
62. *Merismopedia elegans* A. Br.
63. *Merismopedia punctata* Meyen.
64. *Merismopedia tenuissima* Lemm.

PLATE – VI

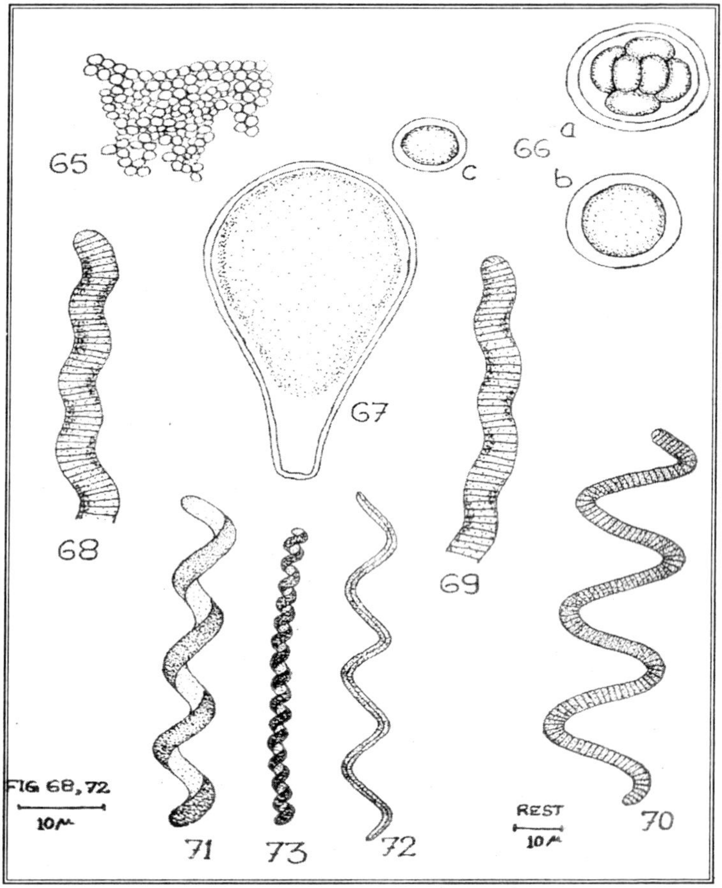

65. *Chlorogloea microcvstoides Geitler.*
66. *Chroococcidiopsis mvsorensis Tiwari, G.S.*
67. *Democarpa olivacea (Reinsch.) Tilden var. gigantea Rao, C.B.*
68. *Arthrosipra balkrishnanii Kamat.*
69. Arthrosipra jenneri Stizenb. Ex Gomont.
70. *Arthrosipra khannae Drouet et Strickland.*
71. *Spirulina gigantea Schmidle.*
72. *Spirulina laxissima West, G.S.*
73. *Spirulina major Kutz. ex Gomont.*

PLATE – VII

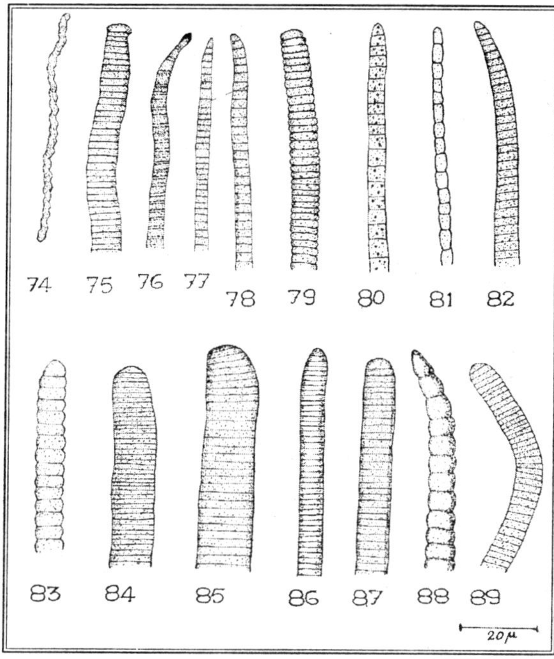

74. *Spirulina subtilissima* Kutz. ex Gomont.
75. *Oscillatoria anquina* (Bory) Gomont.
76. *Oscillatoria amoena* (Kutz.) Gomont.
77. *Oscillatoria animalis* Ag. ex Gomont.
78. *Oscillatoria acuta* Bruhl et Biswas, orth. mut Geitler.
79. *Oscillatoria annae* van Goor.
80. *Oscillatoria amphibia* Ag. ex Gomont.
81. *Oscillatoria amphigranulata* van Goor.
82. *Oscillatoria brevis* (Kutz.) Gomont.
83. *Oscillatoria biswasii* Kamat.
84. *Oscillatoria bharadwajae* Kamat.
85. *Oscillatoria curviceps* Ag. ex Gomont.
86. *Oscillatoria curviceps* var. *angusta* Ghose.
87. *Oscillatoria carakiana* playfair. V. *non-granulata* Kamat.
88. *Oscillatoria cortiana* Meneghini ex Gomont.
89. *Oscillatoria chalybea* (mertens) Gomont var. *insularis* Gardner.

PLATE – VIII

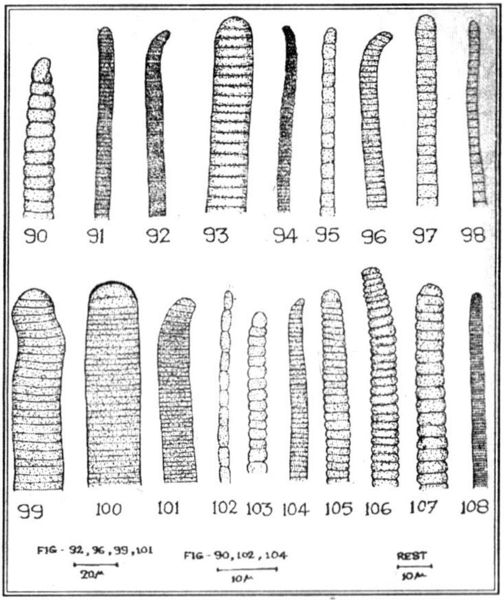

90. *Oscillatoria* Chalybea (Mertens) Gomont V. Minor Kamat.
91. *Oscillatoria* Chilkensis Biswas
92. *Oscillatoria* Chlorina Kutz. Ex Gonont.
93. *Oscillatoria Decoloratea West, G.S.*
94. *Oscillatoria foreaui Fremy.*
95. *Oscillatoriaqeitleriana Elekin.*
96. *Oscillatoria grunowiana Gomont.*
97. *Oscillatoriairriqua (kut.) Gomont.*
98. *OscillatoriaJasorvensis Vouk.*
99. *Oscillatorialimosa iasorvensis Vouk.*
100. *Oscillatoria limosa Ag. Ex Gomont.*
101. *Oscillatoria limosa Ag. Ex Gomont var. disperse-qranulata Schorb.*
102. Oscillatoria limnetica Lemm.
103. Oscillatorialaete-virens (Crouan) Gomont.
104. *Oscillatoria (Crouan) Gomont. Minimus Biswas.*
105. Oscillatoria mouqeotii kutz
106. *Oscillatoria nigroviridis Thwaites ex Gamont.*
107. Oscillatoria Voucher
108. Oscillatoria Bruhl et Biswas.

PLATE – IX

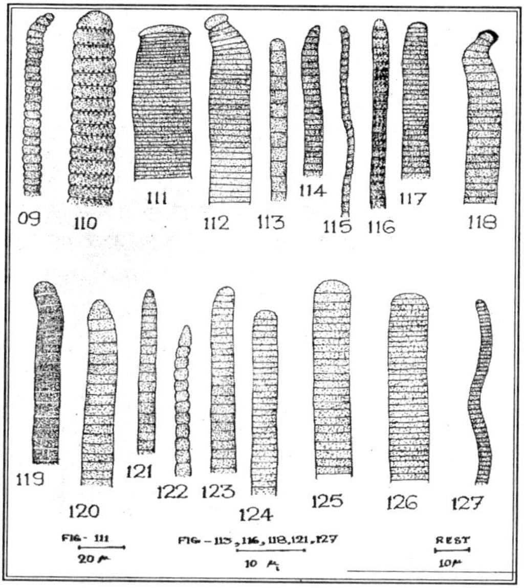

109. Oscillatoria okeni Ag. Ex. Gomont.
110. *Oscillatoria ornata* Kutz. ex. Gomont.
111. *Oscillatoria princeps* Vaucher ex. Gomont.
112. *Oscillatoria proboscidea* Gomont.
113. *Oscillatoria profunda* Krrchner.
114. *Oscillatoria prolifica* (Grev.) Gomont
115. *Oscillatoria pseudogeminata* G. Schmid.
116. *Oscillatoria quadripunctulata* Bruhl et. Biswas.
117. *Oscillatoria rubescens* DC ex. Gomont.
118. *Oscillatoria* rubescens DC ex. Gomont. Var. *Curvata* Kamat
119. *Oscillatoria raoi* De Toni, J.
120. *Oscillatoria raciborskii* Wolosz.
121. *Oscillatoria schultzii* Lemm.
122. *Oscillatoria schulzii* Lemm. V. cylindrical Kamat.
123. *Oscillatoria subuliformis* Kutz. ex. Gomont.
124. *Oscillatoria subbrevis* schmidle.
125. *Oscillatoria subbrevis* schmidle. f. *crassa* Dixit.
126. *Oscillatoria subbrevis* schmidle. f. *major* West, G.S.
127. *Oscillatoria subtilissima Kutz.*

PLATE – X

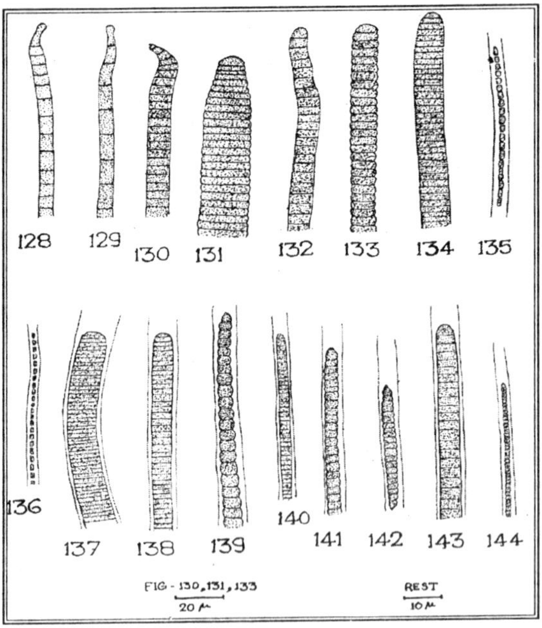

128. *Oscillatoria splendid* Grev. ex. Gomont.
129. *Oscillatoria splendid* Grev. ex. Gomont. Var. *attenuata* W. and G.S. West.
130. *Oscillatoria sancta* Biswas.
131. *Oscillatoria* sancta (Kutz.) Gomont.
132. *Oscillatoria terebriformis* Ag. ex. Gomont.
133. *Oscillatoria tenuis* Ag. ex. Gomont.
134. *Oscillatoria vizagapatensis* Rao, C.B.
135. *Phormidium africanum* Lemm.
136. *Phormidium angustissimum* W. et. G.S. West
137. *Phormidium anomala* Rao, C.B.
138. *Phormidium amiguum* Gomont.
139. *Phormidium abronema Skuja.*
140. *Phormidium bohneri* Schmidle.
141. *Phormidium ceylanicum* Wille.
142. *Phormidium ceylanicum* Wille. Var. minor Kamat.
143. *Phormidium corium* (ag.) Gomont. Var. capitatum Gardner.
144. *Phormidium cebennense* Gomont.

PLATE – XI

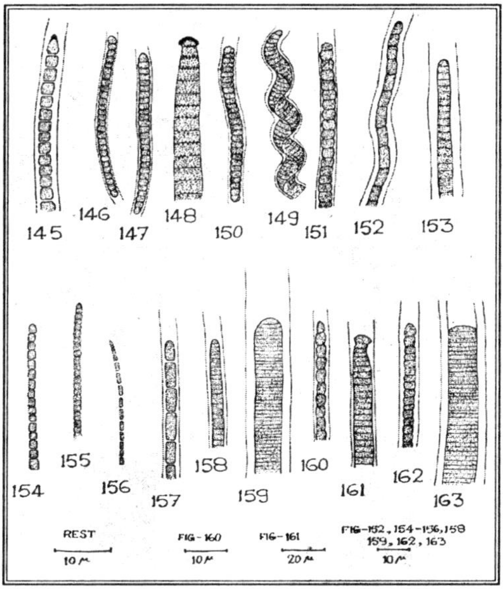

145. *Phormidium dimorphum Lemm.*
146. *Phormidium foveolarum (Mont.) Gomont.*
147. *Phormidium fragile (Meneghini) Gomont.*
148. *Phormidium favosum (Bory) Gomont.*
149. *Phormidium hieronymusii Lemmermann.*
150. *Phormidium jenkelianum Schmid, G.*
151. *Phormidium luridum (Kutz.) Gomont.*
152. *Phormidium mucosum Gardner.*
153. *Phormidium mucosum Gardner. Var. arvense* Rao, C.B.
154. *Phormidium molle (Kutz.) Gomont.*
155. *Phormidium molle (Kutz.) Gomont. F. tenuior W. and G.S. West.*
156. *Phormidium mucicola Hub.– Pestalozzi et Naumann.*
157. *Phormidium orientale West, G.S.*
158. *Phormidium papyraceum (Ag.) Gomont.*
159. *Phormidium stagnina Rao, C.B.*
160. *Phormidium tenue (Menegh.) Gomont.*
161. *Phormidium uncinatum (Ag.) Gomont.*
162. *Lyngbya attenuata Fritsch.*
163. *Lyngbya aestuarii Liebm. ex. Gomont. Var. tenuis Dixit.*

PLATE – XII

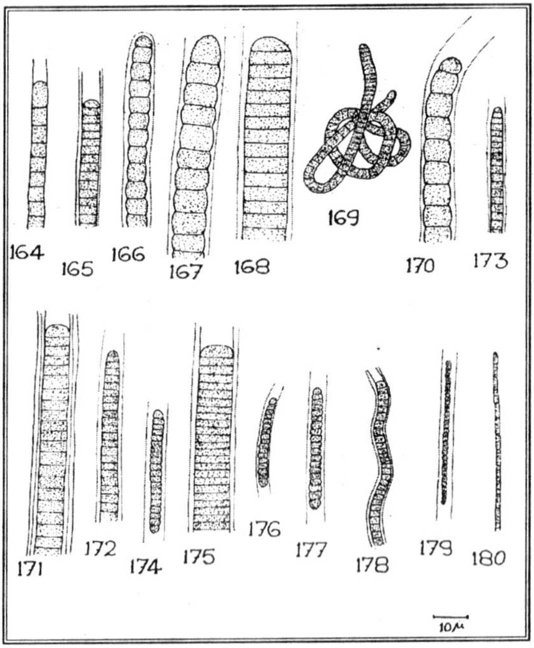

164. *Lyngbya allorgei Fremy*
165. *Lyngbya aerugineo-coerulea (Kutz.) Gomont.*
166. *Lyngbya cryptovaginata Schkorbatow.*
167. *Lyngbya ceylanica Wille.*
168. *Lyngbya corticicola Bruhi et Biswas.*
169. *Lyngbya contorta Lemm.*
170. *Lyngbya dendrobia Bruhl et. Biswas.*
171. *Lyngbya dendrobia Bruhl et. Biswas f. lurida Bruhl et. Biswas Gomont.*
172. Lyngbya dendrobia Bruhl et. Biswas Var. *skujaii nom. nov.*
173. *Lyngbya digueti Gomont.*
174. *Lyngbya gardneri (Setchell et Gardner) Geitler*
175. *Lyngbya hieronmusii Lemm*
176. *Lyngbya Kuetzingii Schmidle.*
177. *Lyngbya kuetzingiana Kirchner.*
178. *Lyngbya lagerheimii (Mob.) Gomont.*
179. *Lyngbya lachneri (zimmermann) Geitler.*
180. *Lyngbya limnetica* Lemmermann.

PLATE – XIII

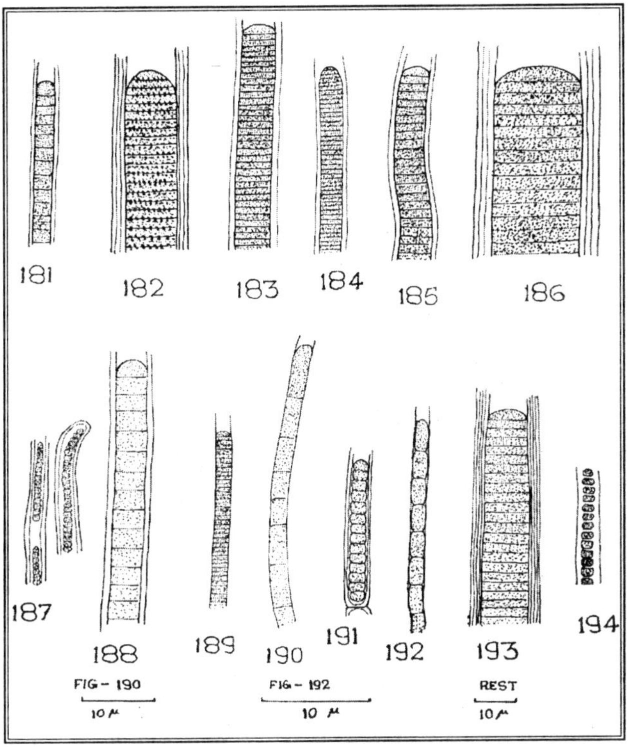

181. *Lyngbya ltea (Ag) Gom.*
182. *Lyngbya major Menegh ex. Gomont.*
183. *Lyngbya martensiana Menegh. ex. Gomont.*
184. *Lyngbya martensiana Menegh. ex. Gomont. Var. minor Gardner.*
185. *Lyngbya maharastrensis Kamat.*
186. *Lyngbya majuscula Harvey ex. Gomont.*
187. *Lyngbya nordgardhii Wille.*
188. *Lyngbya palmarum (Martens) Bruhl et. Biswas.*
189. *Lyngbya polysiphoniae fremy.*
190. *Lyngbya perelegans Lemm.*
191. *Lyngbya putealis Mont. ex. Gomont.*
192. *Lyngbya rivularianum Gom.*
193. *Lyngbya stagnina Kutzing.*
194. *Lyngbya scotti Fritsch.*

PLATE – XIV

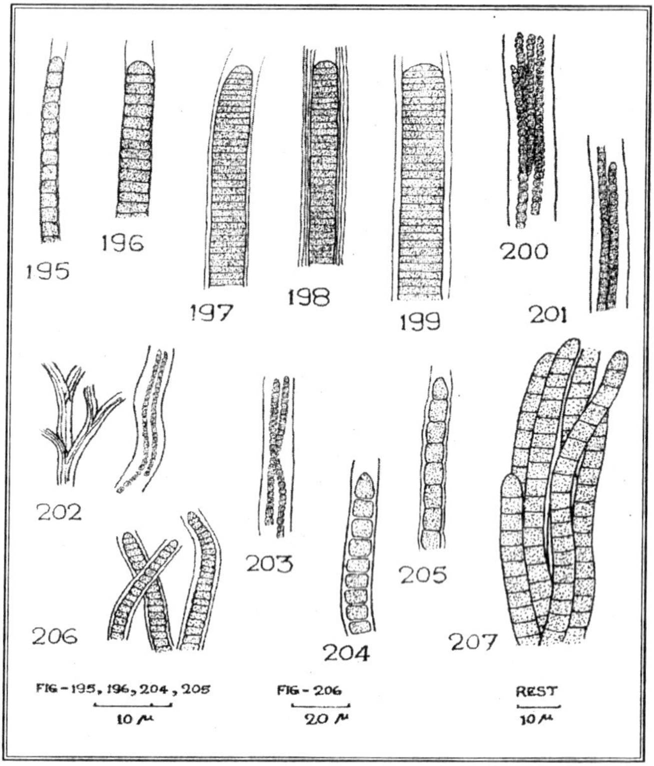

195. *Lyngbya subtilis West, W. Var. granulosa Bruhl and Biswas*
196. *Lyngbya schmidlei V. major Kamat.*
197. *Lyngbya semoplena (C.Ag.) J. Ag. ex. Gomont.*
198. *Lyngbya shackletoni W. et. G.S. West.*
199. *Lyngbya truncicola Ghose*
200. *Schizothrix arenaria (Berk.) Gomont.*
201. *Schizothrix lardacea (Ces.) Gomont.*
202. *Schizothrix lacustris A. Br. ex. Gomont.*
203. *Schizothrix tenuis Woronich.*
204. *Symploca cartilaginea (Mont.) Gomont.*
205. *Symploca elegans Kutzing ex. Gomont.*
206. *Symploca muralis Kutzing ex. Gomont.*
207. Symploca muscorum (Ag.) Gomont.

PLATE – XV

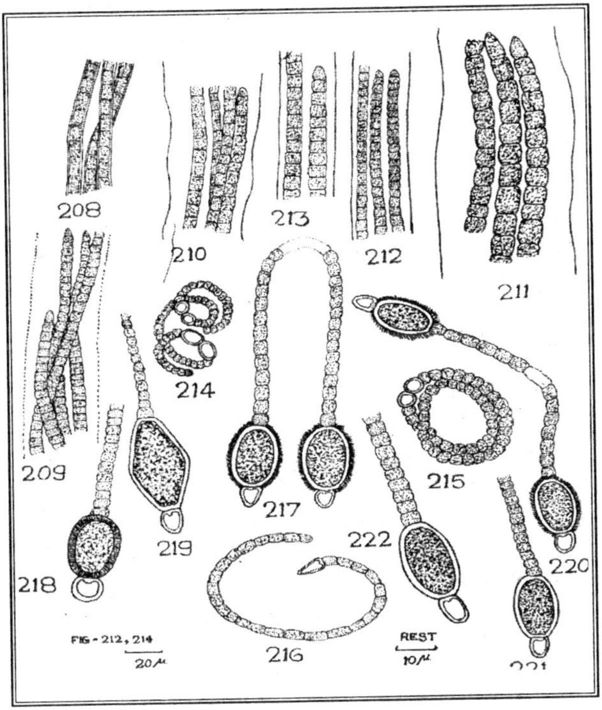

208. *Symploca parietina (A. Br.) Gomont.*
209. *Microcoleus acutissimus Gardner.*
210. *Microcoleus amplus Gardner*
211. *Microcoleus chthonoplastes Thurst ex. Gomont.*
212. *Microcoleus paludosus (Kutz.) Gomont.*
213. *Hydrocoleus subincrustaceus. Hansgrig.*
214. *Anabaenopsis amolodii Aptekarj Var. indica Ramanathan.*
215. Anabaenopsis circularis (G.S. West) *Wolosz. et. Miller*
216. *Anabaenopsis tanganyikae (West, G.S.) Wolosz et. Miller*
217.Cylindrospermum gorakhpurense Singh, *R.N.*
218. *Cylindrospermum indicum Rao, C.B., orth. Mut. De. Toni.*
219. *Cylindrospermum licheniforme Kutz ex. Born. et. Flah.*
220. *Cylindrospermum majus Kutzing ex. Borm et. Flah.*
221. Cylindrospermum musicola Kutzing ex. *Borm et. Flah.*
222. Cylindrospermum musicola Kutzing ex. Borm et. Flah.

PLATE – XVI

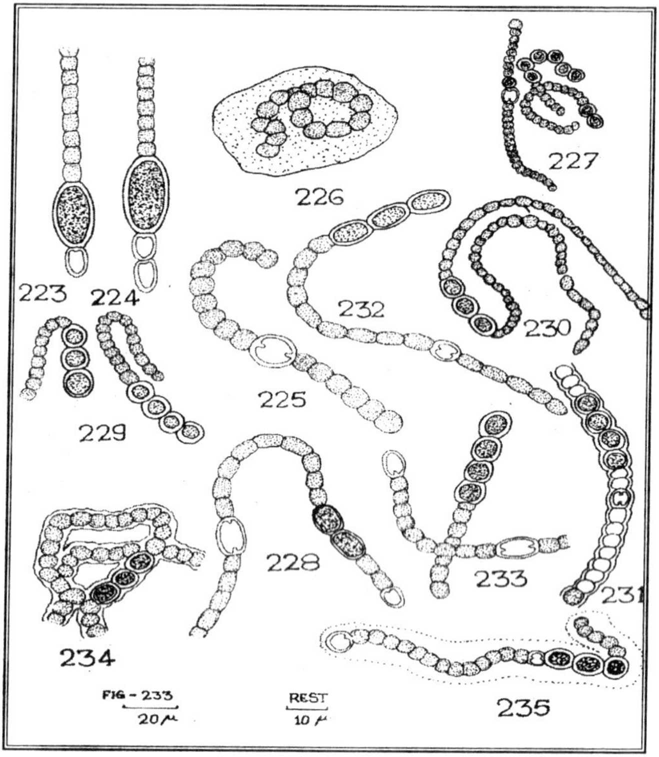

223. Cylindrospermum musicola Kutzing ex. Borm *et. Flah.*
224. Cylindrospermum michailovskoense *Elenkin.*
225. *Nostoc coeruleum Lyngbye ex Born. et. Flah.*
226. *Nostoc commune Vaucher ex Born. et. Flah.*
227. *Nostoc calcicola Brebisson ex Born. et. Flah.*
228. *Nostoc ellipsosporm (Desm.) Rabenh. ex Born. et. Flah.*
229. *Nostoc entophytum Born. et. Flah.*
230. *Nostoc humifusum Carmichael ex Born. et. Flah.*
231. *Nostoc linckia (Roth) Bornet ex Born. et. Flah.*
232. Nostoc microscopicum Carm. ex Born. et. *Flah.*
233. *Nostoc muscorum Ag. ex Born. et. Flah.*
234. *Nostoc maculiforme Born. et. Flah.*
235. Nostoc paludosum Kutzing ex Born. et. Flah.

PLATE – XVII

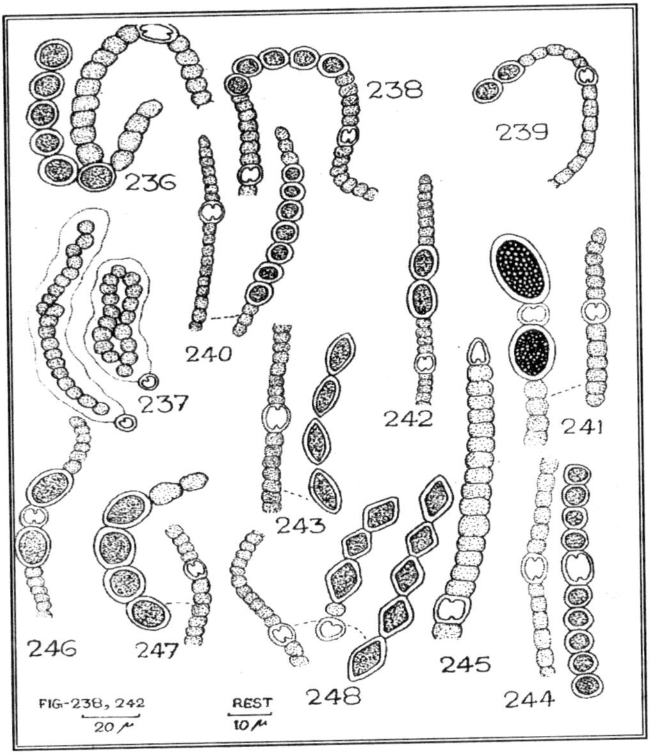

236. Nostoc piscinale Kutzing ex Born. et. Flah.
237. Nostoc punctiforme (Kutz.) Harjot.
238. Nostoc spongiaeforme Agardh ex Born. et. *Flah.*
239. Nostoc sphaericum Vaucher ex Born. et. *Flah.*
240. *Ababaena anomala Fritsch.*
241. *Ababaena ambigua Rao., C.B.*
242. *Ababaena circunalis Rabenhorst ex Born. et. Flah.*
243. *Ababaena doliolum Bharadwaja.*
244. *Ababaena fertilissima Rao, C.B.*
245. *Ababaena Iyengarii Bharadwaja.*
246. *Ababaena Iyengarii Bharadwaja Var. tenuis Rao, C.B.*
247. *Ababaena laxa (Rabehn.)*
248. *Ababaena naviculoides Fritsch.*

PLATE – XVIII

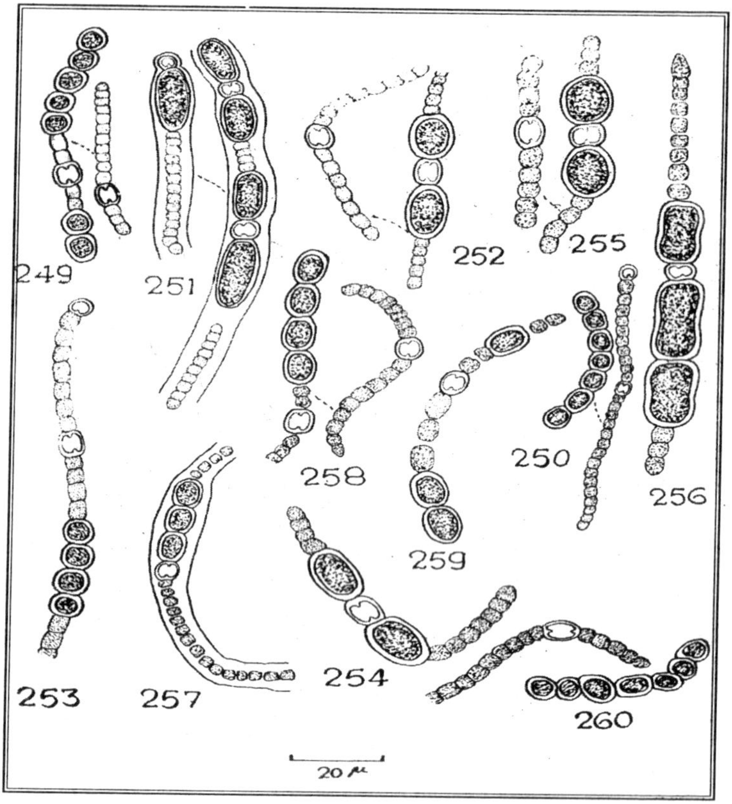

249. *Ababaena oryzae Fritsch.*
250. *Ababaena oscillariodes Bory ex Born. et. Flah.*
251. *Ababaena oscillariodes Bory ex Born. et. Flah. Var. angustus Bharadwaja.*
252. *Ababaena orientalis Dixit.*
253. *Ababaena randhawe Venkataraman, G.S.*
254. *Ababaena sphaerica Bornet et. Flahault.*
255. *Ababaena sphaerica Bornet et. Flahault. Var. attenuata Bharadwaja.*
256. *Ababaena torulosa (Carm.) Lagerh. ex Born. et. Flah.*
257. *Ababaena vaginicola Fristsch et Rich.*
258. *Ababaena variabilis Kutzing ex Born. et. Flah.*
259. *Ababaena variabilis Var. ellipsospora.*
260. *Ababaena variabilis Var. ellipsospora.*

PLATE – XIX

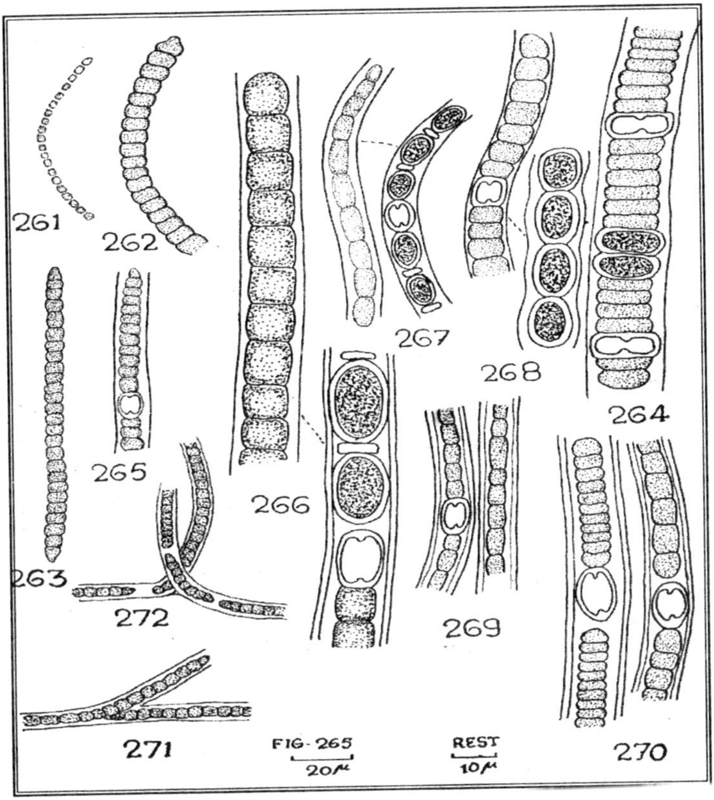

261. *Pseudanabaena catenata Lauterb.*
262. *Pseudanabaena schmidlei Jaag. O.*
263. *Pseudanabaena schmidlei Jaag. O. f. gracilis Skuja*
264. *Nodularia spumigena mertens ex Born. et. Flah.*
265. *Aulosira aenigmatica Fremy.*
266. *Aulosira fertilissima Ghose.*
267. *Aulosira fertilissima Ghose Var. tenuis Rao, C.B.*
268. *Aulosira implexa Bornet et. Flah.*
269. *Aulosira prolific Bharadwaja.*
270. *Aulosira pseudoramosa Bharadwaja.*
271. *Plectonema gracillimum (Zopf) Hansgirg.*
272. *Plectonema hansgirgi* schmidle.

PLATE – XX

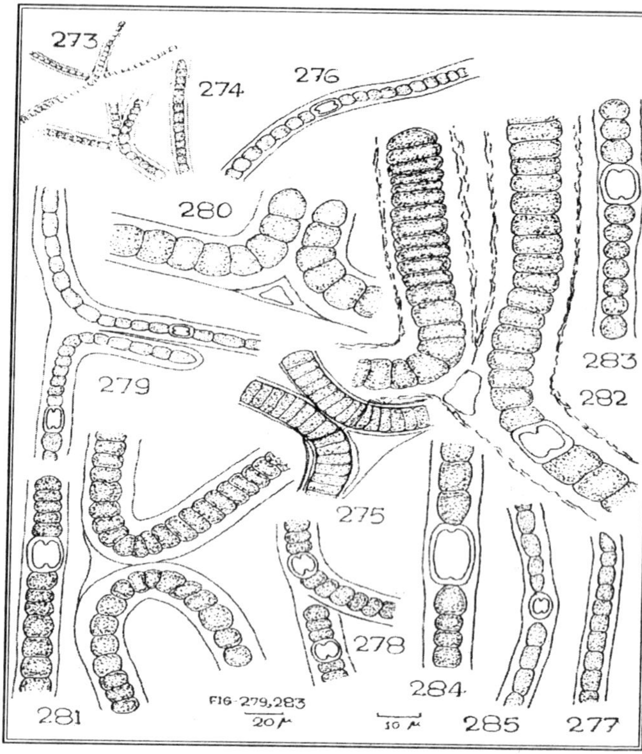

273. *Plectonema nostocorum Bornet. ex Gomont.*
274. *Plectonema puteale (Kirchner) Hansgirg.*
275. *Plectonema tomasinianum (Kutz) Born. et. Gomant.*
276. *Camptylonemopsis iyengarii Desikachary.*
277. *Scytonematopsis woronichinni Kiss.*
278. *Scytonema cincinnatum Thuret ex. Born. et Flah.*
279. *Scytonema fremyii non. nov.*
280. *Scytonema hofmanni Ag. ex. Born. et Flah.*
281. *Scytonema javanicum (Kutz.) Bornet ex. Born. et Flah.*
282. *Scytonema millei Bornet ex. Born. et Flah.*
283. Scytonema ocellatum Lungbye ex. Born. et *Flah.*
284. *Scytonema pseudohofmanni Bharadwaja.*
285. *Scytonema schmidlei J. De. Toni.*

PLATE – XXI

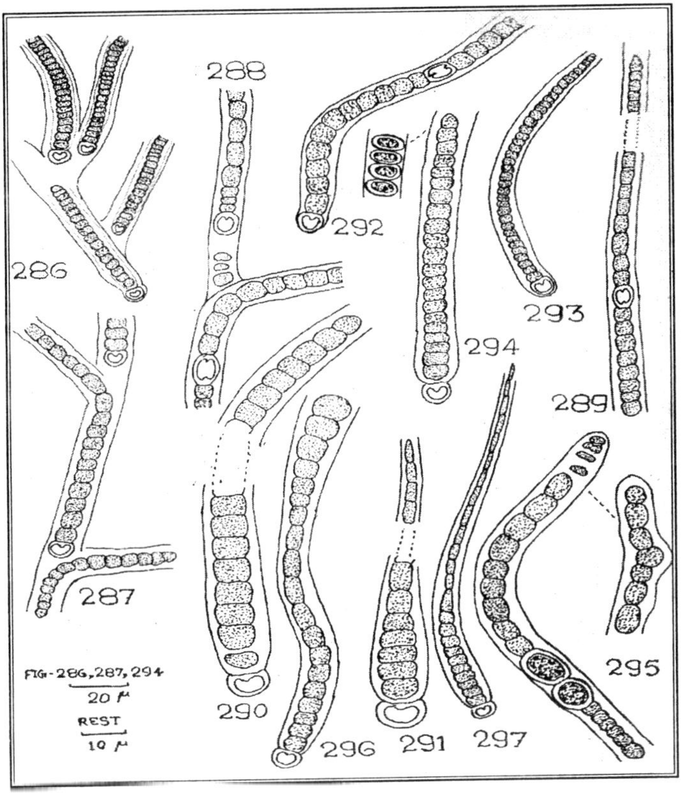

286. *Tolypothrix bouteillei (Berb. Et. Desm.) Forti.*
287. *Tolypothrix byssoidea (Berk.) Kirchner.*
288. *Tolypothrix nodosa Bharadw.*
289. *Microchaete aegualis (Fremy) Comb. nov.*
290. *Microchaete calothrichoides Hansgirg.*
291. *Microchaete loktakensis Bruhi et Biswas.*
292. *Microchaete tenera Thuret ex. Born. et Flah.*
293. *Microchaete tenera Thuret ex. Born. et Flah.*
294. *Microchaete uberrima Carter, N.*
295. *Microchaete uberrima Carter, N. f. minor Carter.*
296. *Fortiea incerta Skuja.*
297. *Calothrix braunii (A.Br.) Bornet. et Flahault.*

PLATE – XXII

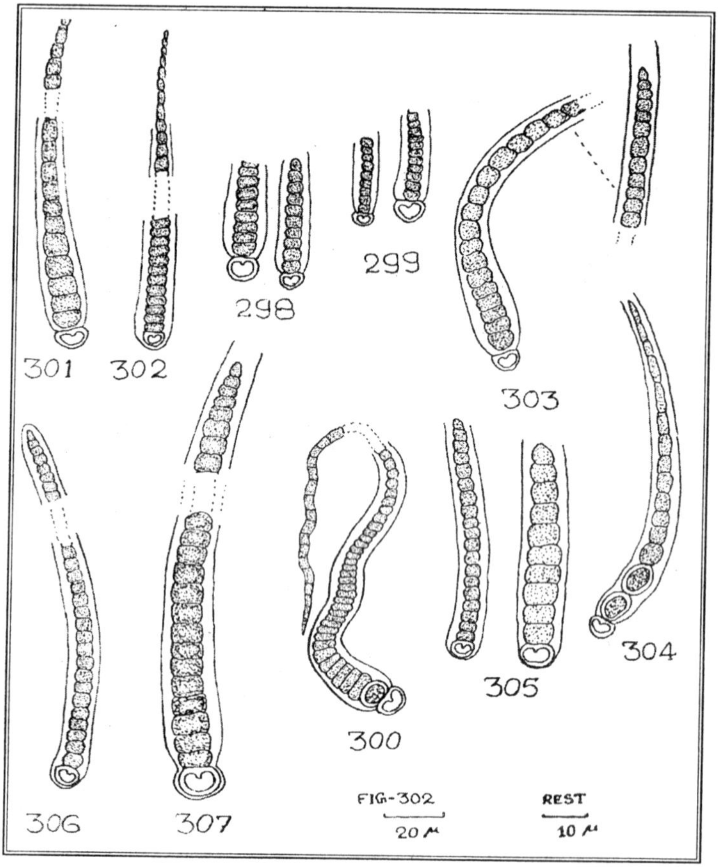

298. *Calothrix brevissima* West, G.S.
299. Calothrix brevissima West, G.S. var. *moniliforme Ghose orth. Mut. Geitler.*
300. *Calothrix castellii (Massal.) Born. et Flah.*
301. *Calothrix clavata* West, G.S.
302. *Calothrix epiphytica W. et West, G.S.*
303. *Calothrix elenkinii* Kossinskaja
304. *Calothrix javanica de Wilde.*
305. *Calothrix membranacea* Schmidle.
306. *Calothrix marchica* Lemmermann.
307. *Calothrix marchica* Lemmermann. Var. *crassa Rao, C.B.*

PLATE – XXIII

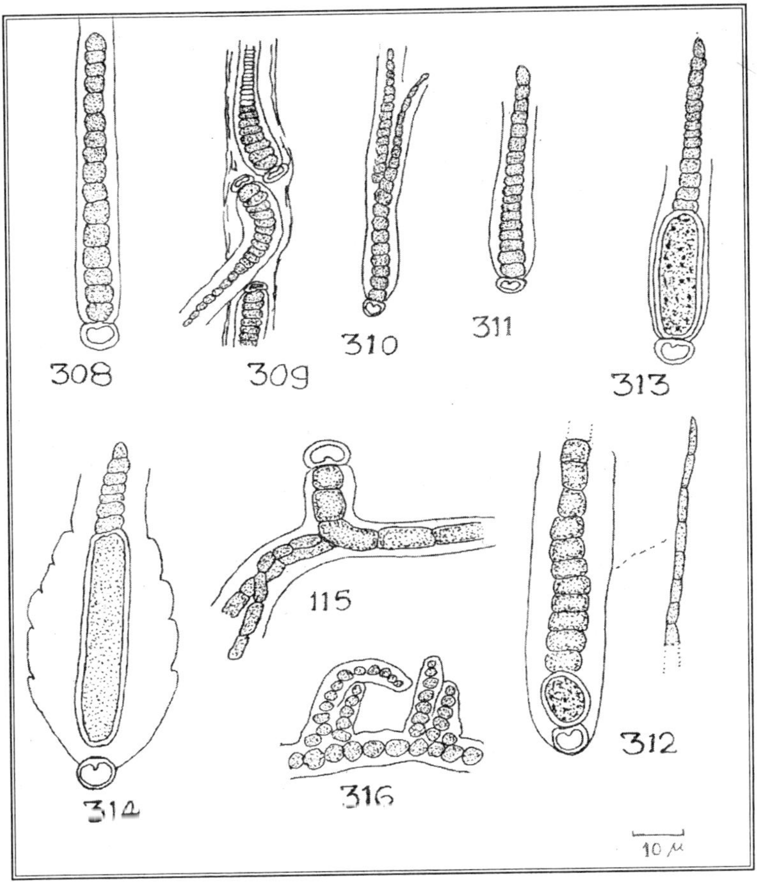

308. *Calothrix marchica* Lemmermann. Var. *intermedia Rao, C.B.*

309. *Dichothrix baueriana* (Grun.) Born. et *Flah.*

310. *Rivularia beccariana (De not.) Born. et Flah.*

311. *Rivularia beccariana (Poir) Berk. ex Born. et Flah.*

312. *Gloeothrichia indica Schmidle.*

313. *Gloeothrichia intermedia (Lemm.) Geitler Var. kanwaensis Rao, C.B.*

314. *Gloeothrichia natans Rabenhorst ex Born. et Flah.*

315. *Nostochopsis lobatus Wood em. Geitler.*

316. *Hapalosiphon baronii W. et. G.S. West*

PLATE – XXIV

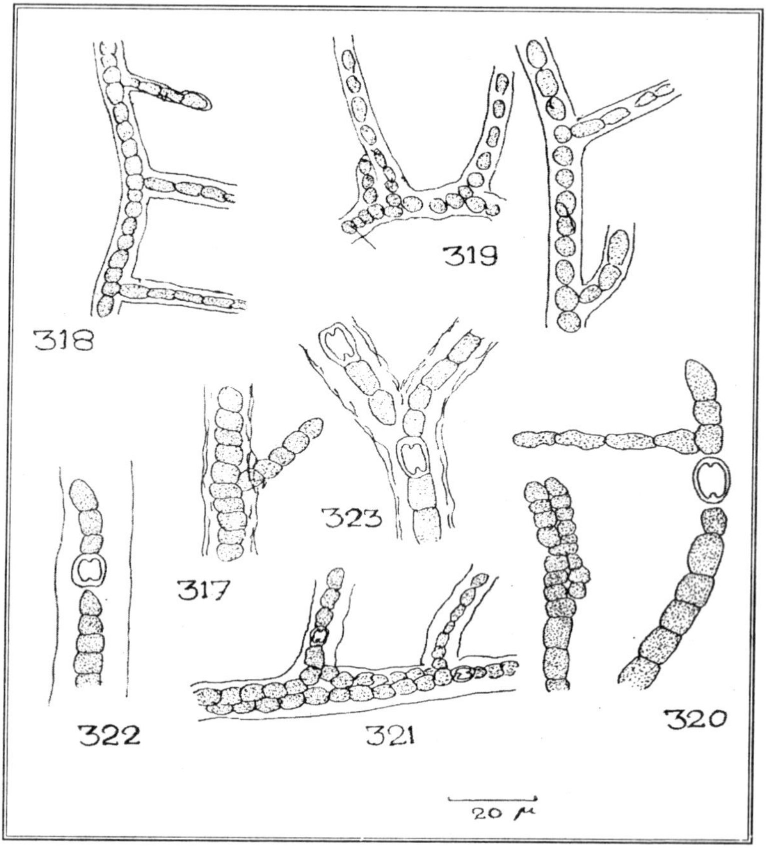

317. *Hapalosiphon hansgirgi (Schmidle) Forti.*
318. *Hapalosiphon intricatus* W. et. *G.S. West.*
319. *Hapalosiphon welwitschii* W. et. *G.S. West.*
320. *Westiellopsis prolifica* Janet.
321. *Fischerella muscicola* (Thuret) *Gom.*
322. *Stigonema hormoides* (Kutz.) *Born, et. Flah.*
323. *Stigonema lavardei* Fremy Var. *indica Fremy.*

PLATE – XXV

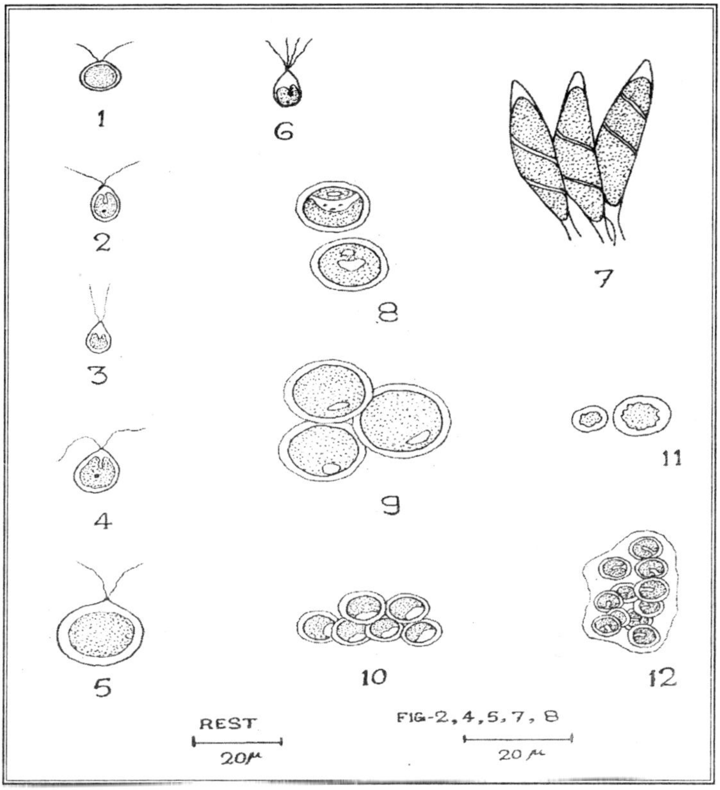

1. *Chlamydomonas globosa* Snow.
2. *Chlamydomonas lismorensis* Playton.
3. *Chlamydomonas mucicola Schmidle.*
4. *Chlamydomonas reinhardi* Dangeard
5. *Chlamydomonas sphaginicola. Fritsch et. Takeda.*
6. *Carteria klebsii* (Dang.) Dill.
7. *Characium ambiguum. Hermann ex Rabnehhorst.*
8. *Chlorococcum humicolo* (Naegeli) Rabnehhorst.
9. *Chlorococcum infusionum (Schrank) Meneghini*
10. *Chlorococcum vitiosum.* Printz.
11. *Trebouxia humicolo (Treboux) West. Et. Fritsch.*
12. *Chlorella conglomerata* (Artari) oltamanns.

PLATE – XXVI

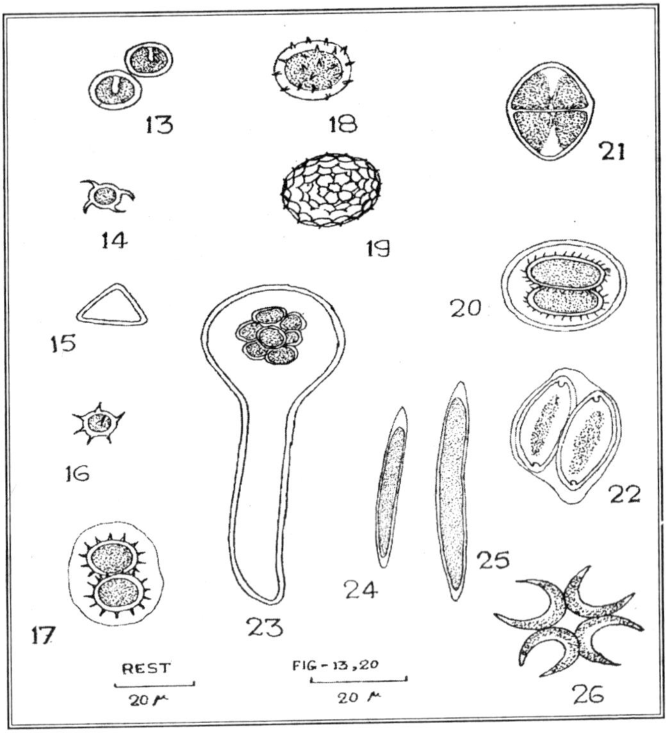

13. *Chlorella vulgaris Beijerinck.*
14. *Tetraedron guadratum (Reinsch) Hansgirg. F. minus. (Reinsch) De Toni.*
15. *Tetraedron muticum (A. Braun.) Hansgirg.*
16. *Tetraedron octaedricum v. spinosum (Reinsch.) W. et. G.S. West.*
17. *Trochiscia aciculifera (Lagerb.) Hansgirg.*
18. *Trochiscia granulata (Reinsch.) Hansgirg.*
19. *Trochiscia reticularis (Reinsch.) Hansgirg.*
20. *Oocystis elliptica W. West.*
21. *Oocystis lacustris. Chodat.*
22. *Oocystis nodulosa West et West.*
23. *Protosiphon botryoides (Kutz.) Kelbs.*
24. *Ankistrodesmus falcatus (Corda) Ralfs.*
25. *Ankistrodesmus falcatus (Corda) Ralfs. V. tumidus. (West et West) G.S. West.*
26. *Ankistrodesmus falcatus (Corda) Ralfs.*

PLATE – XXVII

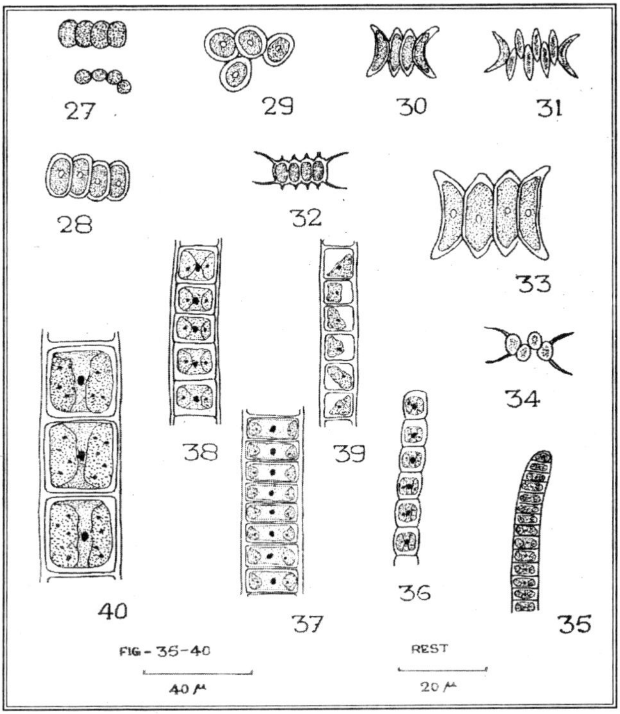

27. *Scenedesmus arcuatus (Lemmermann) Limmermann.*
28. *Scenedesmus bijugatus (Turpin) Kutzing Var. Bicellularis (Chodat) comb. nov.*
29. *Scenedesmus bijugatus (Turpin) Kutzing Var. irregularis (Wille) Smith.*
30. *Scenedesmus dimorphus. (Turpin) Kutzing.*
31. *Scenedesmus falcatus Chodat.*
32. *Scenedesmus longus v. breviseina. Smith.*
33. *Scenedesmus oblicuus. (Turpin) Kuetzing.*
34. *Scenedesmus quadricaude v. maximus W. et. G.S. West.*
35. *Ulothrix oscillarina Kuetzing.*
36. *Ulothrix subtilissima Rabenhorst.*
37. *Ulothrix tenissima Kuetzing.*
38. *Ulothrix tenerrima (Kuetzing) Kuetzing.*
39. *Ulothrix variabilis (Kuetzing) Kuetzing.*
40. *Ulothrix zonata (Weber et Mohr) Kuetzing.*

PLATE – XXVIII

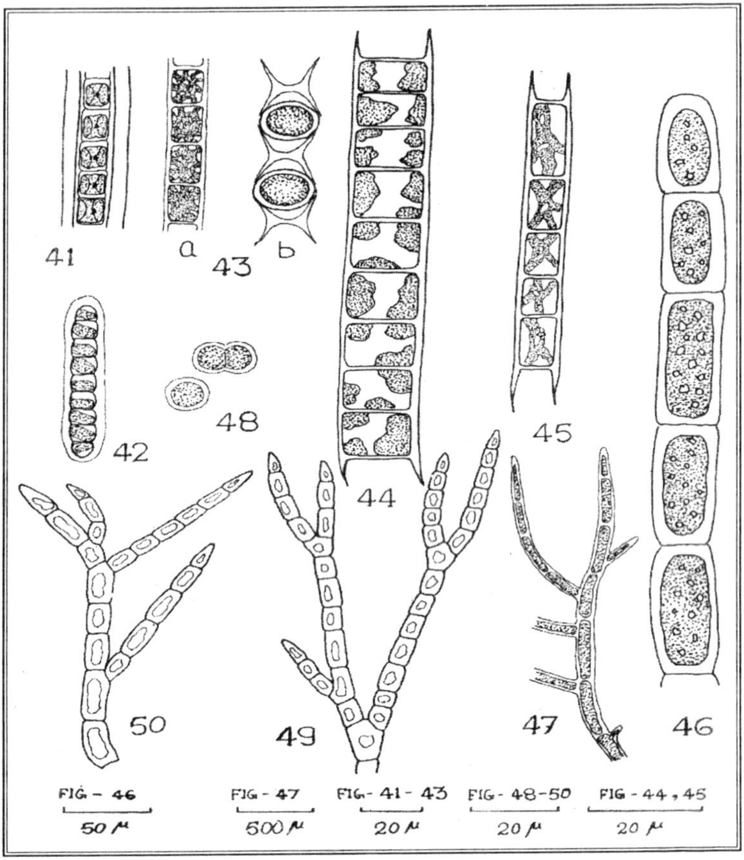

41. *Geminella minor (Naegeli) Heering.*
42. *Geminella protogenita (Kuetzing) G.S. West.*
43. *Microspora abbreviate (Rabenhorst) Lagerheim.*
44. *Microspora floccosa (Vaucher) Thuret.*
45. *Microspora stagnorum (Kutzing) Lagerheim.*
46. *Rhizoclonium heiroglyphicum. (Agardh) Kuetzing.*
47. *Rhizoclonium hookeri Kuetzing.*
48. *Rotococcus viridis Ag. (Prescott) Smith. F.*
49. *Chaetophora elegans (Roth) Ag.*
50. *Chaetophora psisformis Ag.*

PLATE – XXIX

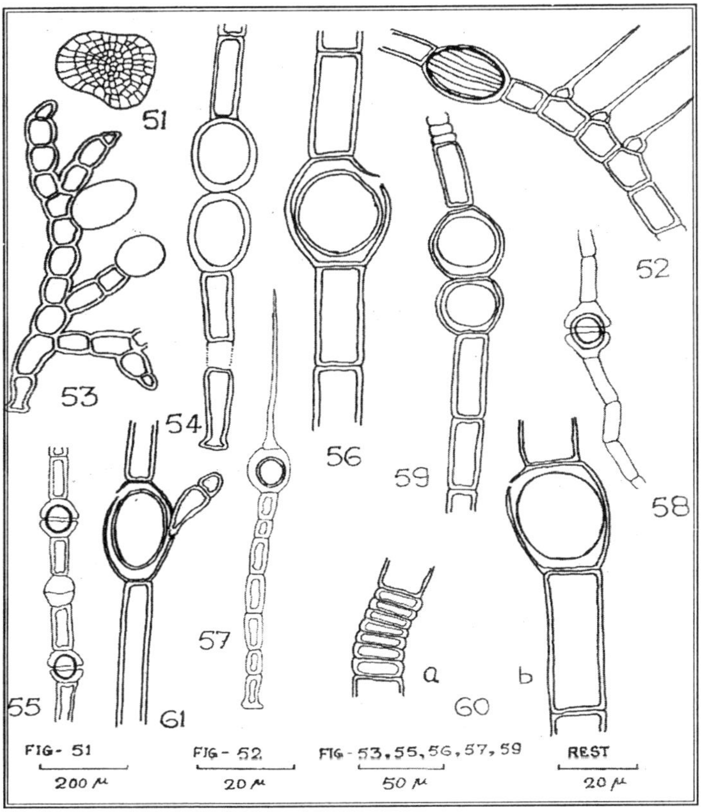

51. *Chaetophora scutata de Brebisson*
52. *Bulbochaete brevifulta Wittrock ex Hirn.*
53. *Oedocladium indicum Kamat.*
54. *Oedocladium aquaticum Kamat.*
55. *Oedocladium formosum Kamat.*
56. *Oedocladium globoum Nordstedt ex. Hirn*
57. *Oedocladium longipilum Jao.*
58. *Oedocladium unisporum Kamat.*
59. *Oedocladium maharastrensis Kamat.*
60. *Oedocladium princeps (Hassall) Wittrock ex. Hirn.*
61. *Oedocladium rugulosum Nordstodt ex Hirn.*

PLATE – XXX

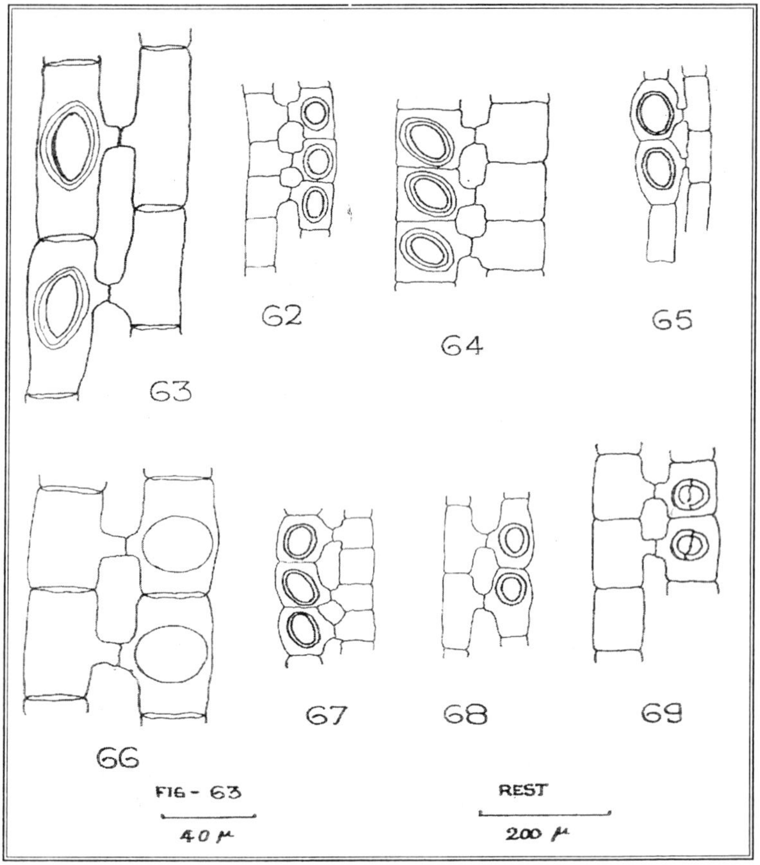

62. *Spirogyra arta Jao.*
63. *Spirogyra biformis Jao.*
64. *Spirogyra communis (Hassal) Kutzing.*
65. *Spirogyra chakiense Kolkwitz and Krieger.*
66. *Spirogyra dictyospora Jao.*
67. *Spirogyra ghosei Singh.*
68. *Spirogyra hyaline Cleve.*
69. *Spirogyra irregularis Nageli.*

PLATE – XXXI

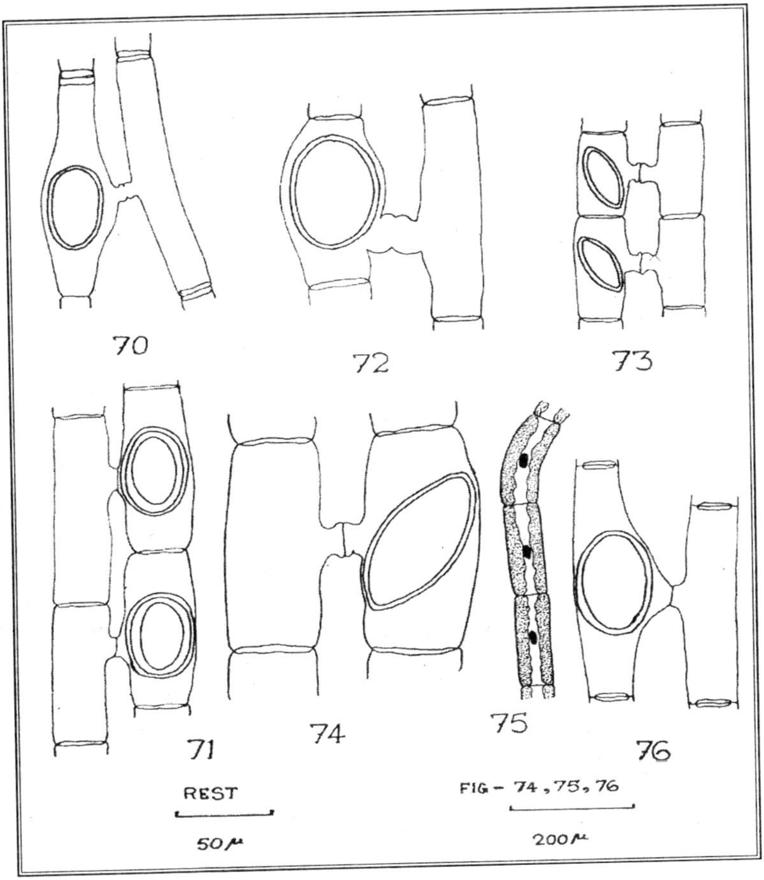

70. *Spirogyra inflata (Vaucher) Kuetzing.*
71. *Spirogyra maxima (Hassall) Wittrock.*
72. *Spirogyra prolifica kamat.*
73. *Spirogyra submaxima Transeau.*
74. *Spirogyra turfosa Gay.*
75. *Sirocladium maharashtranse Randhawa.*
76. *Sirocladium ventersicum Transeau.*

PLATE – XXXII

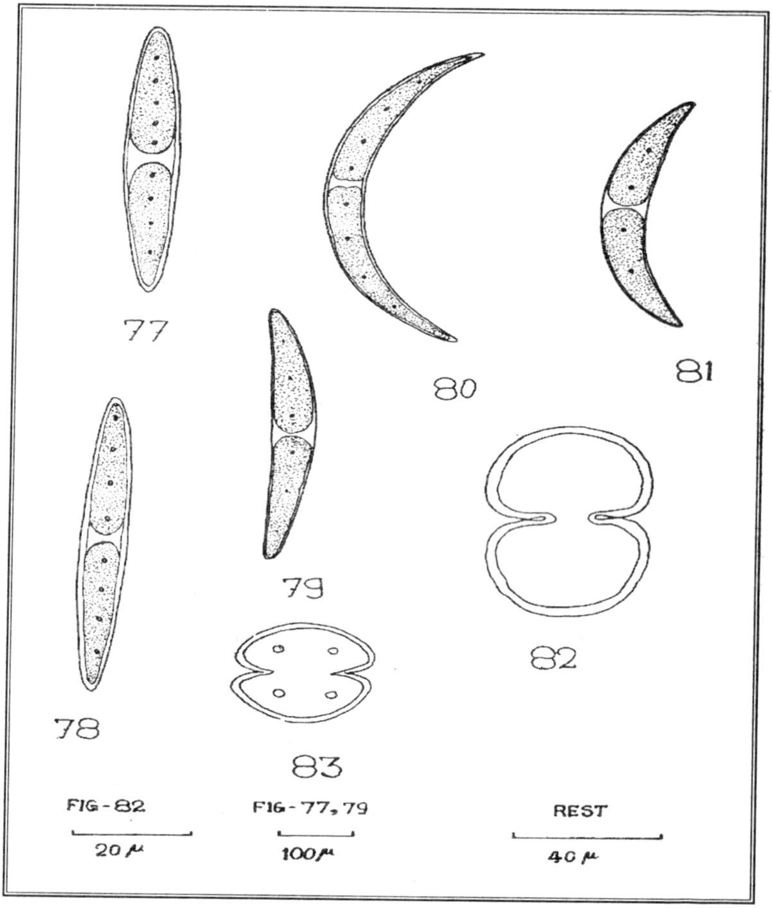

77. *Closterium acerosum (Schrank) Ehrenberg.*
78. *Closterium acutum (Lyngbye) de Brebisson.*
79. *Closterium langeolatum Kuetzing.*
80. *Closterium paryulum Nag.*
81. *Closterium venus Kuetzing.*
82. *Cosmarium contractum Kirchner.*
83. *Cosmarium depressum (Naeg.) Lund. Var. Achondrum (Boldt) West et. West.*

PLATE – XXXIII

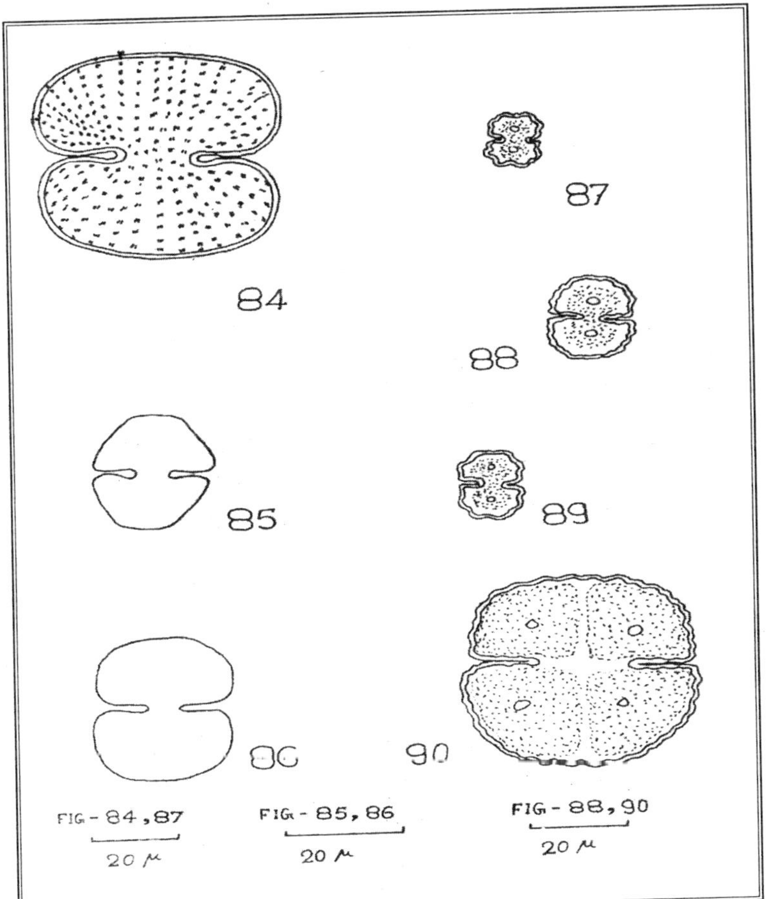

84. *Cosmarium dentiforum Corda Var. minus Kamat.*
85. *Cosmarium granatum Brebisson.*
86. *Cosmarium impressulum Elfv.*
87. *Cosmarium meneghinii de Brebisson.*
88. *Cosmarium subimpressulum. Borge.*
89. *Cosmarium trilobatum Reinsch.*
90. *Cosmarium undulatum Corda ex Ralfs.*

PLATE – XXXIV

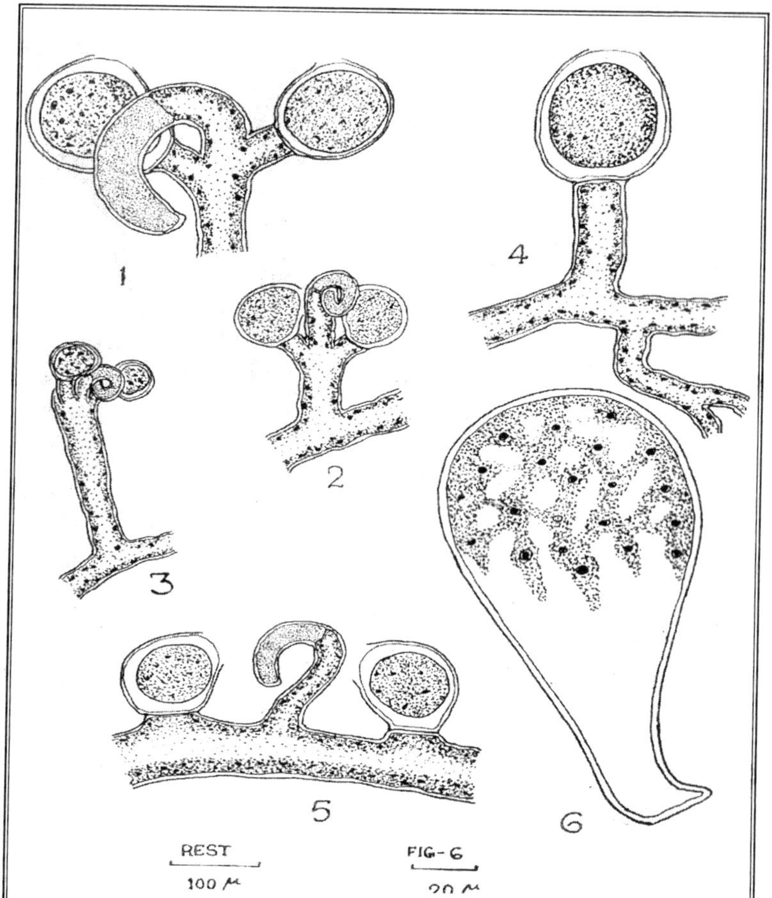

91. *Vaucheria amphibian Randhawa.*
92. *Vaucheria geminata (Vauch.) De Condolle.*
93. *Vaucheria hamata Walz.*
94. *Vaucheria mayyanadensis Erady.*
95. *Vaucheria sessilis (Vauch) De Condolle.*
96. *Botrydium species.*

PLATE – XXXV

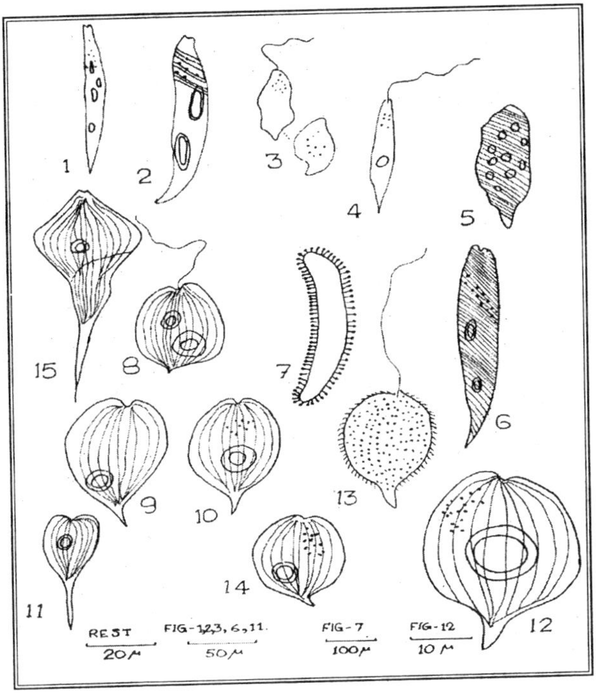

97. *Euglena acus Ehrenberg v. hyalina Klebs.*
98. *Euglena charkowiensis Swir.*
99. *Euglena gracillis Klebs.*
100. *Euglena multiformis Schiller.*
101. *Euglena tuberculata Swirenko.*
102. *Euglena fusca (Klebs) Lemm.*
103. *Euglena maharastrensis Kamat.*
104. *Phacus acuminatus stokes v.discifera (Pochm) Huber-Pestalozzi.*
105. *Phacus caudatus Huebner.*
106. *Phacus hamatus Pochmann.*
107. *Phacus longicauda (E) Dujardin.*
108. *Phacus platalea Drez. v. minor kamath.*
109. *Phacus horridus Pochmann.*
110. *Phacus curvicauda Swirenko.*
111. *Phacus tortus (Lemm.) Skvorkzov.*

PLATE – XXXVI

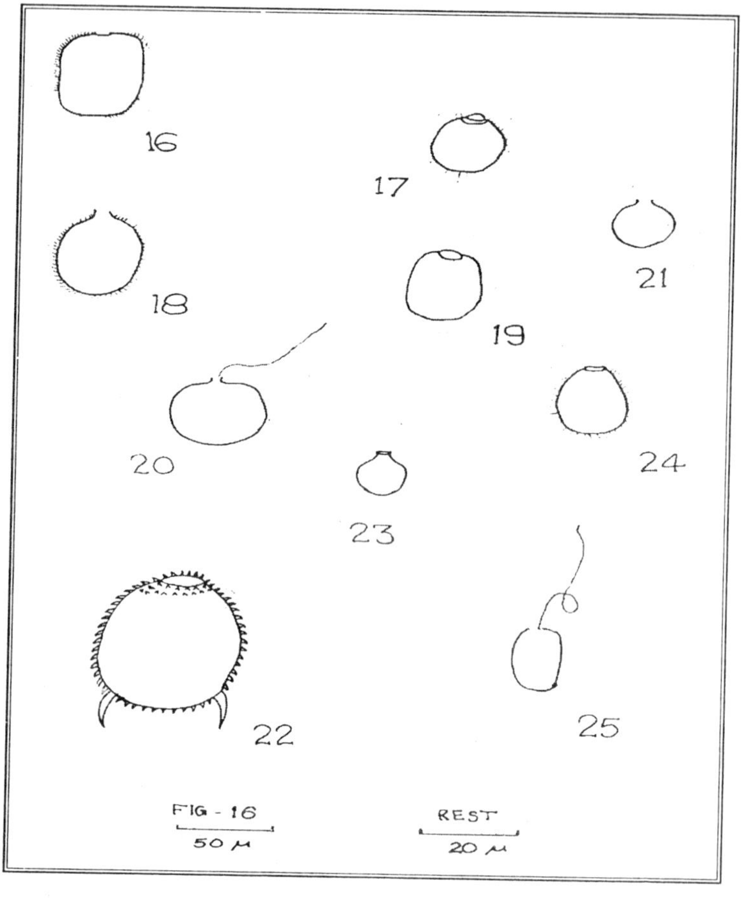

112. *Trachelomonas allia Drez. em. Deflandre.*
113. *Trachelomonas hispida (Perty) Stein emnd. Deflandre.*
114. *Trachelomonas lefevrei Deflandre.*
115. *Trachelomonas ablonga Lemmermann.*
116. *Trachelomonas volvocina Ehrenberg.*
117. *Trachelomonas volvocina Ehrenberg v. derephora Conrad.*
118. *Trachelomonas armata (E) stein v. steinii emend. Deflandre.*
119. *Trachelomonas cordata (Drez.) comb. Deflandre f. minor Deflandre.*
120. *Trachelomonas hispida (Perty) Stein emend. Deflandre f. minor Bourr.*
121. *Trachelomonas pulcherrima Playfair.*

PLATE – XXXVII

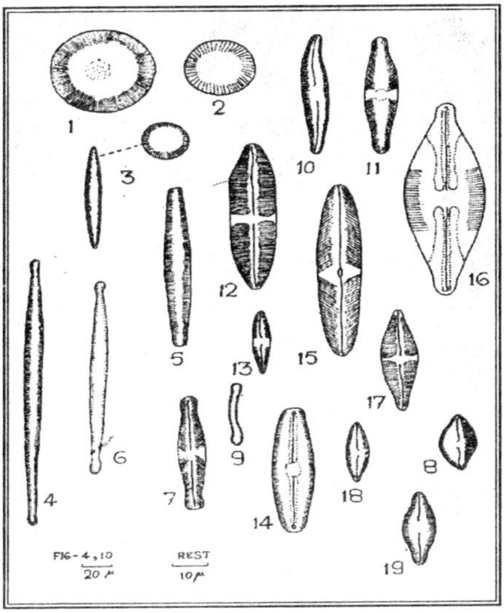

1. *Cyclotella kutzingiana Thwaites.*
2. *Cyclotella operculata Kg.*
3. *Fragillaria brevistriata f. elongata Venk.*
4. *Synedra affinis. Kg.*
5. *Synedra pulchella var. lanceolata o. Meara.*
6. *Synedra ulna (Nitzsch.) Ehrenb.*
7. *Achnanthes coarctata Berb.*
8. *Achnanthes delicatula Kg.*
9. *Achnanthes microcephala. Kg.*
10. *Gyrosigma spenceri (W.Sm.) Cl.*
11. *Stauroneis anceps Ehrenh.*
12. *Navicular anceps (Ehrenb.) var. linearis Ehrenb. (als Art).*
13. *Navicula aponina Kg.*
14. *Navicula bacilliformis Grun.*
15. *Navicula brebissonii Kg.*
16. *Navicula clavata Greg.*
17. *Navicula crucicula (W. Sm.) Donk.*
18. *Navicula cryptocephala Kg. f. terrestris Lund.*
19. *Navicula gregaria Dunk.*
20. *Navicula grivillei (Ag.) Cl.*

PLATE – XXXVIII

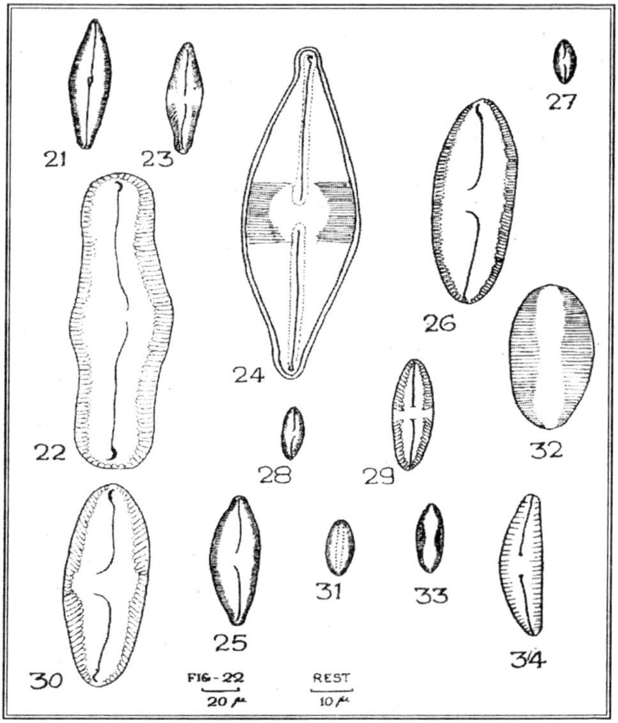

21. *Navicula lanceolata (Ag.) Kg.*
22. *Navicula major Kg.*
23. *Navicula mutica Kutz.*
24. *Navicula sculpta Ehernb.*
25. *Navicula salinarum Grun.*
26. *Navicula terrestris Petersen.*
27. *Navicula tennelloides (Hust.) emend.*
28. *Navicula ucheriae var. densistriata Lund.*
29. *Pinnularia microstauron (Ehr.) Cleve.*
30. *Pinnularia viridis (Nitzsch.) Ehr.*
31. *Amphora caffaeiformis Ag.*
32. *Amphora ovalis Kg.*
33. *Amphora Montana (Krasake).*
34. *Cymbella alpine Grun.*

PLATE – XXXIX

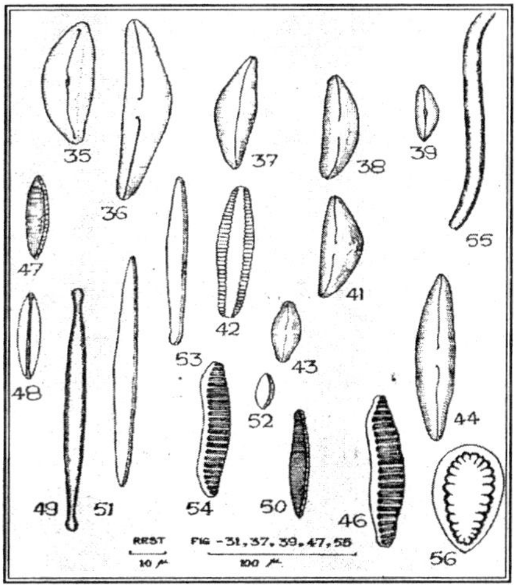

35. *Cymbella austriaca Grun.*
36. *Cymbella cymbiformis (Kg.) V.H.*
37. *Cymbella gracilis (Rebenh.) Cleve.*
38. *Cymbella parva W. Sm*
39. *Cymbella reinhardtii Grun.*
40. *Cymbella tumescens. A. Cl.*
41. *Cymbella ventricosa (kg.) Cleve.*
42. *Rhopalodia gibba (Ehrenb.) O. Muller Var. Ventricosa (Enrenb.) Grun.*
43. *Gomphonema angustatum (Kg.) Grun. var. prodjcta Grun.*
44. *Gomphonema gracile Ehernb.*
45. *Gomphonema lanceolatum Ehr.*
46. *Hantzschia anphioxys Grun.*
47. *Nitzschia angustata (W.Sm.) Grun..*
48. *Nitzschia dissipata (Kg.) Grun.*
49. *Nitzschia gracilis Hantzsch.*
50. *Nitzschia hantzschiana Rabenh.*
51. *Nitzschia intermedia Hantzsch.*
52. *Nitzschia kuetzingiana Hilse.*
53. *Nitzschia palea (Kg.) W.Sm.*
54. *Nitzschia parvula lewis var. terricola Lund.*
55. *Nitzschia vermicularis (Kg.) Hantzsch.*
56. *Surirella baltica Schum.*

SYSTEMATIC ACCOUNT OF THE ALGAE

Division	-	Cyanophyta
Class	-	Cyanophyceae
Order	-	Chroococcales
Family	-	Chroococcaceae
Genus	-	*Microcystis* Kutzing

Cells spherical or elongated, many in spherical, ellipsoidal or irregularly overlapping or net-like colony, free-swimming, often with attached daughter colonies; cells in homogeneous colorless, often diffluent, mucilage, individual envelopes absent; cells mostly very densely arranged, cell-division in all directions, generally transverse in elongate cells; gas-vaculoes often present; nannocytes present in M. flos-aquae.

1. Microcystis elabens (Breb.) Kutz.

Plate-I, Fig. 1
Colony spherical or flat and expanding, blue-green or olive-green when old together with number of daughter colonies; cells oblong, 2-4.5 μ broad, (3-) 4-8.5 μ long, with gas-vacuoles.
In cultures.

2. Microcystis holsatica Lemmermann

Plate- I, Fig. 2
Colonies spherical, clathrate, margins of colonial mucilage well defined; cells closely arranged, 1 μ diam., spherical or subspherical, with gas-vacuoles, blue-green.
In cultures as well as perceived aligned to the paddy under water.

3. Microcystis insignis (Bech) Comb. nov.

Plate- I, Fig.3

Colonies distinct, macroscopic, 3 mm diam., globose, more or less diffluent; cells spherical or subspherica1, 3.7 - 4.9 μ lonq, 3.7 μ broad, pale blue-green.

Perceived in water bodies of paddy field.

4. Microcystis marginata (Menegh.) Kutz.

Plate- I, Fig. 4

Colony round or irregularly flattened and more or less lenticular, generally simples; margins of colonial mucilage very distinct, refractive, sometimes stratified; single colony ellipsoidal to avoid in out line, 140 - 150 (-200) μ in length and 60-95 μ in width; cells 3-6 μ in diam., closely arranged, with gas-vacuoles.

Perceived aligned to the plant portions under water.

5. Microcystis protocystis Crow.

Plate- I , Fig. 5

Colonies irregular, often diffuse, with limits of colonial mucilage not clearly delimited, sometimes disappearing cells very numerous, varying in the mode of aggregation from closely packed to generally dissociated, spherical, 3.5-6.5 μ diam., with gas-vacuoles.

Perceived in water bodies of paddy field

6. Microcystis pulvera Var. incerta (Lemm.) Crow.

Plate - I, Fig. 6

Colonies generally, elongated upto 4 times often many together, limits of colonial mucilage distinct; cells spherical or ellipsoidal, closely arranged, 1-2 μ diam., blue-green or olive-green without gas-vacuoles.

In moist soil as well as in cultures.

7. Microcystis robusta (Clark) Nygaard.

Plate- I , Fig. 7 .
Colonies at first round, later irregularly elongate and clathrate; sheath distinct, later gelatinizing; cells 6-9 μ diam., spherical, without gas-vacuoles.
In cultures.

8. Microcystis stagnalis Lemm.

Plate- I, Fig. 8
Colonies very long and small, sometimes expanding and clathrate, colonial mucilage indistinct; cells very closely arranged, spherical, 1-2 μ diam., pale blue-green, without gas-vacuoles.
Perceived in water bodies of paddy fields.

9. Macrocystis scripta (Richter) Lemm.

Plate - I , Fig. 9
Colonies more or less elongate, like cuneate figures, often clathrate, while young round, ring like, colonial mucilage distinct; cells 4.5-7 μ, seldom up to 8 μ broad with gas-vacuoles.
In cultures.

Genus - **Chroococcus** Nag.

Cells spherical or subspherical, hemispherical after division, in small groups of 2-4 individuals, sometimes 8-16, rarely single, in a gelatinous or mucous matrix; sheath of individual cells distinct, firm, generally lamellated, in some homogeneous, persistently or irregularly broken; reproduction by cell division and fragmentation of colonies; division of cells in 3 directions. Nannocytes occasionally seen.

10. Chroococcus granulosus Zeller.

Plate - I, Fig. 10

Thallus gelatinous, orange coloured; cells 4-12 rarely many together, colonies about 25 μ diam., cells 4.5 - 7 μ diam. when single about 15 μ diam., spherical or angular; sheath thin, hyaline when young not conspicuous, later broad.

In moist soil and in cultures.

11. Chroococcus limneticus Lemm.

Plate-I. Fig.II

Cells spherical or subspherical after division, 4-32, free floating, mostly in a tubular gelatinous layer, without sheath cells 6-12 μ diam, with sheath 8-14 μ diam., sheath distinct or diffluent, unlamellated, colourless, colonial mucilage broad, cell division generally in two, seldom in three directions, cell contents gray, blue-green, olive-green or yellowish.

In cultures.

12. Chroococcus macrococcus (Kutz.) Rabenh.

Plate - I, Fig .12

Thallus mucilaginous, somewhat broad, yellowish brown, more or less dilated; cells spherical, 2-4 together, also single, 25-50 μ diam., with sheath 30-80 μ diam., sheath thick, colourless, lamellated.

In cultures.

13. Chroococcus minutus. (Kutz.) Nag.

Plate - I, Fig. 13

Cells spherical or oblong, single or in groups of 2-4, light blue-green, with sheath 6-15 μ diam., and without sheath 4-10 μ diam; colonies 10-13 x 15-20 μ sheath not lamellated, colourless.

In green patches with other algae.

14. Chroococcus montanus Hansgirg.

Plate-I, Fig.14

Thallus slimy gelatinous, brownish or blackish brown; cells 5-6 µ, seldom 3-9 µ diam., single or 2-4, seldom more in irregular groups of 19-25 µ diam., rarely up to µ diam., colonial sheath lamellose.

In cultures.

15. Chroococcus minor (Kutz.) Nag.

Plate - I, Fig. 15

Thallus slimy-gelatinous, dirty blue-green, or olive green; cells spherical, 3-4 µ in diam., singly or in pairs, seldom 4 or 8; sheath colourless, very thin, hardly visible.

In cultures.

16. Chroococcus pallidus Nag.

Plate - II , Fig. 16

Thallus gelatinous, yellowish or colourless; cells single or 2-4, seldom up to 8 in elliptic oblong colonies, without sheath 5-8µ, with sheath 7-11.5 µ (13 u) broad, blue green or yellow; sheath colourless, unlamellated.

On moist soil.

17. Chroococcus schizodermaticus West.

Plate-II, Fig.17

Cells in groups of 2-4, blue-green, without sheath 5.8-11 µ with sheath 21-42 µ diam., sheath yellow to brown, very distinct, lamellated, 5-10 lamellae, outer layers often broken.

In small water bodies.

18. Chroococcus spelaeus Ercegovic.

Plate - II, Fig.18.
Cells solitary or 2-4 together in a colony, globose, blue-green, sheath hyaline, unstratified, cells without sheath 14-16 μ in diameters, with sheath 20-25 μ in diameter.
In cultures.

19. Chroococcus turgidus (Kutz.) Nag.

Plate - II, Fig.19.
Cells spherical or ellipsoidal single, or in groups of mostly 2-4, very seldom many, blue-green, olive green or yellowish, without sheath 8-32 μ, with sheath 13-25 μ diam. rarely 40 μ sheath colourless not distinctly Iamellated.
In cultures.

20. Chroococcus turgidus (Kutz.) Nag. Var. **fuscescens** (Kutz.). Forti.

Plate - II, Fig.20.
Cells spherical single, or in groups of mostly 2-4, very seldom many, blue-green, olive-green or yellowish, without sheath 8-32 μ, with sheath 13-25 μ diam. rarely 40 μ sheath colourless, not distincrly lamellated; protoplasm brownish.
In cultures.

21. Chroococcus tenax, (kirchn.) Hieron

Plate-II, Fig.21
Cells mostly in groups of 2-4, sometimes 8-16, blue-green or olive coloured, without sheath 16-21 μ, with sheath 20-26 μ in diam., sheath colourless or yellow to brown, very thick, very distinctly lamellated, 3-4 lamellae.
In water bodies aligned to paddy stalks.

22. Chroococcus westii (W.West) Boye-Petersen

Plate - I I , Fig. 22

Cells single or in groups of 2-4, without sheath 13-27 μ diam., with sheath 18-32 μ diam., violet; sheath colourless, distictly lamellated, coloured yellow to brownish with chlor-zinc-iodide.

Perceived in algal patches on much wet soils.

Genus - **Gloeocapsa** Kutzing.

Cells spherical 2-8 in colonies, seldom many, with a number of concentric special envelopes; colonies single or many together forming an expanded mass, individual sheaths lamellated or unlamellated, cell division vary regularly in three directions, cells in large colonies often with secondary colonies, arranged irregularly; occasionally with nannocytes, resembling Aphanocapsa stage; spores with firm thick walls often formed in a number of species.

23. Gloeocapsa atrata (Turp.) Kutz.

Plate - II, Fig. 23

Thallus crustaceous, mucilagenous, blackish; cells without sheath 3.5 - 4.5 (-5) μ diem., with sheath 9 - 14.5 μ diam., pale blue-green, mostly many in a colony; sheath colourless or pale bluish, thick unlamellated or indistinctly lamellated.

In wet soil surfaces.

24. Gloeocapsa decorticans (A.Sr.) Richter.

Plate - II, Fig. 24

Cells spherical or sometimes oval, blue-green, single or up to 2-4 together; single cells with 19 x 21 μ without sheath 6 x 8 μ, in a two celled stage with sheath 22 x 30 μ without sheath up to 12 μ long; sheath cloudless, thick distinctly lamelleted.

In cultures.

25. Gloeocapsa montana Kutz.

Plate - II , Fig. 25

Thallus amorphous, broad, mucilaginous, pale yellowish, or light green in colour; cells spherical or sub spherical, with sheath 4 - 10 μ broad and without sheath 2 - 5 μ broad, single or two together in a colony, colony 13 - 28 μ broad; sheath lamellated, colourless, outer lamellae diffluent contents more or less opaque, homogeneous, finely granular, pale blue-green.

In cultures.

26. Gloeocapsa nigrescens Nag.

Plate - II, Fig. 26

Thallus crustaceous, thin, blackish, cells spherical, without sheath 3.3 - 6.8 μ diam., and with sheath 12 - 13.5 μ diam., in colonies of 30 125 μ diam., uniting laterally; sheath broad, not lamellated, light purple, sometimes intense purple or colourless; contents blue green; spores with a smooth thin membrane, 5-7 μ in diam., may in a homogenous sheath.

In cultures.

27. Gloeocapsa polydermatica Kutz.

Plate-II, Fig.27

Thallus mucilaginous, compact; cells spherical, without sheath 3 - 4.5 μ diam., blue-green; sheath colourless very thick, as thick as protoplast, very distinctly and many times lamellated.

On moist soils as well as in cultures.

28. Gloeocapsa punctata Nag.

Plate -II, Fig.28

Thallus gelatinous, light blue-green; cells without sheath 0.7 - 1.5 (-2.8) μ diam., with 3.5 - 7 μ broad, blue green; sheath

thick, colourless, unlamellated or scarcely lamellated; cells 2 - 16 in groups or colonies, about 25 µ diam.
In cultures.

29. Gloeocapsa pleurocapsoides Novacek

Plate-II, Fig.29
Colony microscopically small, more or less roundish or irregular, dull olive green, brownish or blackish brown; cells olive coloured or pale blue green, homogeneous, granular; cells 5 - 10 µ in diam., elongated or angular, closely arranged; sheath thin, firm, simple, coloured, yellowish brown or brownish in the peripheral part 2 - 3 µ thick and lamellated 3 - 4 lamellae.
In cultures.

30. Gloeocapsa rupestris Kutz.

Plate-III, Fig.3D
Thallus brownish, crustaceous; cells without sheath 6 - 9 (-11) µ in diam., blue-green; sheath yellow to brown, outer daughter colonies often pale yellow to nearly colourless very distinctly lamellated; colonies 15 - 75 µ in diam., spores with firm thin blackish brown wall, more or less 15 µ diam. On moist soil surfaces.

31. Gloeocapsa stegophila (Itzigs.) Rebenh.

Plate-III, Fig.31
Cells without sheath 3-4.7 µ, with sheath 4.5-8 µ diam., 4-32 cells in globose groups, 50-140 µ diam; blue-green sheath golden to arrange, yellow to red, mostly unlamellated; spores with a dull reddish brown, thin firm membrane.
In cultures.

Genus - **Gloeothece** Nag

Cells cylindrical to ellipsoidal, straight or bent, not attenuated at the ends, but broadly rounded, in small colonies or forming large thallus, division of cells at right angles to the longitudinal axis, sometimes the daughter cells divide in all the three directions: colonies single or many together forming an expanded mass, individual sheath lamellated or unlamellated, cells in larger colonies arranged irregularly. Nannocytes present.

32. Gloeothece membranacea (Rabenh.) Barnet

Plate - III, Fig. 32

Thallus expanded, leathery, gelatinous, olive-green to brown, about 1 mm in diam., cells without sheath 4.5-5.5 µ broad and 7.5-8.8 µ long, blue-green; sheath colourless, distinct or diffluent.

In cultures.

33. Gloeothece palea (Kutz.) Rabenh.

Plate - III, Fig.33.

Thallus mucilaginous, mostly blue-green; cells long cylindrical, without envelope 2.5-4.5 µ broad, to 3 times as long as broad, with envelope 8-12 µ broad blue-green or nearly colourless; individual mucilaginous envelopes colourless or partly coloured yellowish, not lamellated, nannocytes present.

Perceived aligned to paddy stalks under water.

34. Gloeothece rupestris (Lyngb.) Barnet.

Plate - III, Fig.34.

Cells ellipsoidal to cylindrical without envelope 4-5.5 (-6) µ broad, 1.5-3 times as long as broad, with envelope 8-12 µ broad, contents mostly blue green, 2-4 rarely 8 together in oval to subglobose colonies, colonies 25-41 µ diam., envelopes colourless or brownish at the peripheries, lamellated or unlamellated.

On much wet soils.

35. Gloeothece samoensis Wille.

Plate-III, Fig.35

Cells ellipsoidal, without sheath 4 - 5 μ broad, 8 μ long cells yellowish or bluish green in round colonies, often many uniting, mostly 2 - 4 in a common envelope, envelope colourless, unlamellated.

In cultures.

36. Gloeothece samoensis Wille Var. **Major** Wille

Plate-III, Fig.36

Cells without envelope 5 - 6 μ broad and 8-10 μ long.

In cultures.

Genus - **Aphanocapsa** Nag.

Cells spherical or nearly so, many loosely arranged without an order, forming a formless gelatinous mass, often a few cm in diam., mucilage homogeneous, colourless, cells often with thin more or less gelatinized individual sheaths; division in two (or various) directions, often two, 4 and sometimes many within a common mucilaginous envelope of the parent cell.

37. Aphanocapsa biformis A.Br.

Plate-III, Fig.37

Thallus olive green, gelatinous often expanding; cells 4 - 7 μ diam, spherical, mostly with a special envelope, loosely arranged, 2 - 4 together in a common mucilaginous envelope; nanocytes about 2 μ diam.

In cultures as well as on moist soil.

38. Aphanocapsa banarcsensis Bharadwaja.

Plate - III, Fig. 38

Plant mass soft, spherical, hollow, irregularly hyaline or cream coloured up to 1.5 cm in diam., cells oval or almost spherical, 4 - 6.2 μ in diam., sheath thick, unstratified, hyaline, closely adpressed to the cells, up to1 μ thick.

In cultures.

39. Aphanocapsa delicatissima W. et. G.S. West

Plate - III, fig. 3 9

Colony spherical, ellipsoidal or irregular, 14-30 x 15-50 μ with colourless or yellowish, homogeneous, mucilaginous envelope; cells spherical, O.S - 0.75 μ diam., loosely arranged gray to pale blue-green.

Perceived in water bodies in paddy fields.

40. Aphanocapsa elachista W. et. G.S. West.

Plate - III, Fig.40.

Colony small, spherical or ellipsoidal, 26-38 μ in diam., mucilage thin, colourless, homogeneous, often diffluent, cells very loosely arranged single or in pairs, 1.5-1.8 μ seldom upto 2 μ broad, blue green.

Perceived aligned to paddy surface under water.

41. Aphanocapsa fonticola Hansgirg.

Plate - III, Fig.41.

Thallus dark brown to dark blue-green cells arranged in twos, cell contents blue green, cells spherical, 3.0-3.5 μ in diam.

In cultures.

42. Aphanocapsa grevillei (Hass.) Rabenh.

Plate - III, Fig.42.

Thallus gelatinous spherical or hemispherical, light blue-green; cells spherical, 3.2-5.6 µ diam., contents finely granular, blue-green, closely arranged in a homogeneous mucilage; individual envelopes not distinct.
Perceived in water bodies of paddy fields.

43. Aphanocapsa montana Cramer.

Plate - IV, Fig.43.

Thallus of no definite shape, gelatinous, yellow-green or blue-green, olive coloured light violet or colourless; cells 2.5-4 µ diam., spherical, light blue-green or yellowish, single or in pairs; mucilage colourless, diffluent.
In cultures.

44. Aphanocapsa muscicola (Menegh.) Wille.

Plate - IV, Fig.44.

Colony microscopic; cells spherical, 2-3 µ diam. blue-green, 2-4 together, daughter cells often together in a common mucilage envelope, mucilage thick, colourless, white or greyish blue.
In collections of water bodies of paddy fields.

45. Aphanocapsa nivalis *Lagerheim.*

Plate - IV, Fig.45.

Colony small, microscopic, mucilage homogeneous, olive green, cells spherical. 6 -7 µ in diameter, blue-green loosely arranged.
In cultures.

46. Aphanocapsa pulchra (Kutz.) Rabenh.

Plate - IV, Fig.46.

Thallus gelatinous, homogeneous, blue-green, tuberculate attached or free; cells spherical, 3.5 - 4.5 (-4.8) µ diam., loosely

arranged, single or in two pale blue green; individual sheaths of cells indistinct.
On wet soil surface with other algae.

47. Aphanocapsa roeseana de Bary

Plate-IV, Fig.47
Thallus broad, irregularly spherical, brownish green to blue-green, often gelatinous and up to 30 cm, cells 5 - 8 µ diam., nearly oval, pale blue-green, mucilage sheath homogeneous.
In cultures.

Genus - **Aphanothece**. Nag.

Cells ellipsoidal to cylindrical, straight or slightly bent, many in a more or less shapeless expanded thallus, mucilage homogeneous, occasionally with lamellated individual envelopes, often gelatinizing, division transverse. Nannocytes present.

48. Aphanothece clathrata W. et. G.S. West.

Plate-IV, Fig.48
Thallus microscopic, small, when planktonic, or large when benthic, often irregular, clathrate, with colourless homogeneous mucilage; cells 0.6 - 0.7 (-1) µ broad, (2.5-) 3.5 - 4.5 µ long, cylindrical, straight or slightly curved, densely packed, pale blue-green.
Perceived floating in water bodies of paddy field.

49. Aphanothece conferta Richter.

Plate-IV, Fig.49
Thallus gelatinous, membranous, expanded, dirty green or olive brown; cells single or in twos; oblong, or spherical, 2.5 - 3 µ diam., 1-2 times as long as broad; sheath diffluent, colourless; contents pale blue-green, granular.
In cultures as well as on moist soil surface.

50. Aphanothece castagnei (Breb.) Rabenh.

Plate-IV, Fig.50

Thallus gelatinous, without any definite shape, slimy, blue-green or brown; cells ellipsoidal to cylindrical, 2-3.6 μ broad, 4-8 μ long, mostly densely arranged, blue-green; sheath diffluent, colourless or yellowish.

Perceived floating in water bodies of paddy fields.

51. Aphanothece microscopica Nag.

Plate-IV, Fig.5l

Thallus small, gelatinous, at first round later amorphous, up to 2 mm diam., cells oblong, cylindrical, more or less 4.5 μ broad, 1 ½ -2 times as long as broad, with some- times distinct, individual sheath, blue-green, colourless. Nannocytes present.

In cultures.

52. Aphanothece naeqelii Wartm.

Plate - V, Fig. 52

Thallus gelatinous, yellow-brown or olive green; cells oval, after division spherical, 3.5 - 4.5 x 6.5 - 8 μ in size, blue-green; sheath diffluent.

In cultures.

53. Aphanothece nidulans Richter, P.

Plate-V, Fig.53

Thallus irregularly expanded, often found between other algae, plankton forms, more or less round; cells cylindrical, straight or slightly bent, 1 - 1.5 μ broad, upto 3.5 μ long, blue-green, most densely arranged; mucilage sheath mostly diffluent, colourless or yellow to brownish yellow.

In cultures.

54. Aphanothece pallida (Kutz.) Rabenh.

Plate-V, fig.54

Thallus gelatinous soft, blue-green or brownish, upto 4-6 mm diam., cells oblong, elliptical, or cylindrical, 3-8 µ broad and 1-3 times as long as broad, blue-green, olive green; sheath very distinct in the peripheral part of the thallus, often lamellated, yellow to brown, diffluent in the inner part.

On moist soil surface.

55. Aphanothece stagnina (Spreng.) A.Br.

Plate-V, fig. 55

Thallus gelatinous, spherical, ellipsoidal, up to many cm in diam., pale blue-green, dull brown or brownish, in the inside often with calcareous crystals; cells oblong, more or less avoid or cylindrical, 3 - 6.5 µ broad, 4.5 - 11 µ long, more or less blue-green, densely or sparsely arranged, generally densely in the peripheral region of the colony and sparsely in the inside of the colony, without individual envelopes, homogeneous mucilage.

In cultures.

56. Aphanothece saxicola Nag.

Plate-V, Fig.56

Thallus mucilaginous, colourless, or yellowish; cells cylindrical, 1 - 2 µ broad and 2-3 times as long, single or in pairs, seldom many in a common, mucilaginous envelope, pale blue-green.

In cultures.

Genus - **Synechococcus** Nag.

Cells oblong, cylindrical or ellipsoidal, erect, seldom slightly bent with rounded apices, single or in colonies of 2, rarely in fours, mucilage envelope absent or a very thin and narrow one present; division transverse.

57. Synechococcus aeruginosus Nag.

Plate-V, Fig.57
Cells cylindrical, 5 - 16 μ broad, up to 30 μ long, single, or 2 - 4 together, pale blue-green.
On moist soil surfaces.

58. Synechococcus cedrorum Sauvageau

Plate-V, Fig.58
Cells single or two together, elongate, ellipsoidal (sometimes cylindrical), and finely rounded, 3 - 4 μ broad, 1 ¼ -2 times as long as broad, 5 - 10 μ long, contents blue-green.
In cultures.

59. Synechococcus elongatus Nag.

Plate-V, Fig.59
Cells cylindrical, 1.4 - 2 μ broad, 1-3 times as long as broad, single or 2 - 4 cells together; contents homogeneous and light blue-green.
In cultures as well as in water bodies of paddy fields.

Genus - **Synechocystis** Sauvageau

Cells spherical, single or two together after division or rarely in colonies of a few cells, without distinguishable mucilage envelopes.

60. Synechocystis aquatilis Sauv.

Plate-V, Fig.60
Cells spherical, single or twos, 5-6 μ broad, pale blue-green.
In water bodies of paddy fields.

61. Synechocystis pevalekii Ercegovic.

Plate-V, Fig.61

Thallus indefinite, among other algae; cells spherical, after division hemispherical, 2.5 – 3.5 μ broad, single or two together, contents blue – green homogeneous.
Perceived aligned to wet surfaces of paddy.

Genus - **Merismopedia** Meyen

Cells 4-16 (or more) in tabular colonies arranged in a homogeneous mucilage, generally in fours, arranged in a single plan, free floating, cells globose, oblong before cell division, subspherical after division; contents blue-green rarely violaceous or reddish; multiplication by division in two directions perpendicular to the plane of the colony.

62. Merismopedia elegans A.Sr.

Plate-V, Fig.62
Colonies small or big, 16-4000 celled, cells spherical or oblong, more or less closely arranged, 5 - 7 μ broad, and (5-) 6 - 9 μ long; light blue.
In collections of water from paddy fields.

63. Merismopedia punctata Meyen

Plate V, Fig.63
Colonies small, 4 - 64 cells, about 60 μ broad; cells not closely packed, spher1cal or avo1d, 2.5 - 3.5 μ broad, pale blue-green.
In cultures.

64. Merismopedia tenuissima Lemm.

Plate-V, Fig.64
Cells pale blue-green closely packed in colonies of 16-100 cells, subspherical, 1.3 - 2 μ broad, sometimes individual cells with distinct mucilaginous envelopes.

Family - Entophysalidaceae
Genus - **Chlorogloea** Wille

Cells spherical or ellipsoidal mostly without individual envelopes or with a thin unlamellated sheath; in a common mucilage, in straight erect or radial rows, rows sometimes indistinct forming more or less hemispherical or flat irregularly lobed thalli, sometimes with daughter colonies; cell division in three directions but generally in a single determined direction; gonidia and nannocytes present.

65. Chlorogloea microcystoides Geitler

Plate-Vi, Fig.65

Thallus thin, gelatinous, forming a slightly lobed dull green to brownish crust mostly hemispherical or larger thallus made up by the union of a number of daughter colonies; cells spherical, or ellipsoidal, mostly very closely arranged and polygonal in erect or radial rows of branched or more or less indistinct rows, generally without a distinct individual sheath; colonial mucilage colourless, or yellowish to brownish only in peripheral parts; cells 2 - 3.8 μ dia., blue-green, olive green or yellowish; nannocytes formed in somewhat broader cells by repeated divisions, 1.5 - 2 μ diam.

In cultures.

Order - Chamaesiphonales
Family - Cyanidiaceae
Genus - **Chroococcidiopsis** Geitler.

Cells more or less spherical with a narrow firm membrane, single, or in groups becoming transformed into sporangia; cells formed by the formation of endospores, by the repeated division of the entire contents and the gelatinization of the mother wall; vegetative division absent; endospore formation by division in three directions, up to 32 endospores.

66. Chroococcidiopsis mysorensis Tiwari, G.S.

Plate-VI, Fig.66
Cells spherical with a firm membrane, 9.0 - 14.3 μ in diam., endospores present; perennating cells 16 - 25 μ in diam., yellowish, thick walled.
In cultures.

Family - Dermocarpaceae Geitler.
Genus - **Dermocarea Crouan**
Cells spherical, club-shaped, oval, seldom hemispherical, attached, single or gregarious and closely packed, by transverse division a small stalk cell is cut off, mostly without a stalk or with a small mucilage stalk; endospores 4- to many by division in three directions, entire contents involved in endospore formation or a small stalk is differentiated in the beginning, liberated by the gelatiniza- tion of the sporangial wall.

67. Democarpa olivacea (Reinsch) Tilden Var. **gigantea** Rao, C.B

Plate-VI, Fig.67
Plants single, broadly club-shaped, contracted at the base to form an elongated stalk; sporangia with pseudovagina 45 - 56 μ broad and 70 - 100 μ long, without pseudovagina 20 - 43 μ broad and 33 - 76 μ long; family stratified; endospores numerous.
In cultures.
Family - Oscillatoriaceae Kirchner
Genus - **Arthrospira** Stizenberger

Trichomes multicellular, cylindrical, without sheath, loosely and regularly coiled, usually of relatively large diameter and large spirals, with comparatively short and fewer coils; cross-walls distinct, apices slightly or not all tapering, terminal cell rounded, calyptra absent.

68. Arthrospira balkrishnanii Kamat

Plate-VI, Fig.68

Trichomes long with regular spirals, not tapering towards the apices, not constricted at the cross-walls; cells shorter than broad, 3.5 - 4 (-5) μ in diam., 0.8 - 1.2 μ long; spirals 6 - 7 μ broad, distance between the two consequtive spirals 12 - 15 u: end cell rounded without a cap or calyptra.

In cultures.

69. Arthrospira jenneri Stizenb. Ex Gomont

Plate-VI, Fig.59

Thallus blue-green; trichomes blue-green, unconstricted at the cross-walls or only very little constricted 5 – 8 μ broad, not attenuated at the ends, more or less regularly spirally coiled, spirals 9 - 15 μ broad, distance between two spirals 21 - 31 μ cells nearly as long as broad, or somewhat shorter than broad, 4 - 5 long, the cross-walls granulated; end cells broadly rounded.

In water bodies of paddy fields.

70. Arthrospira khannae Drouet et Strickland

Plate-VI, Fig.70

Trichomes planktonic, blue-green, forming loose spirals, spirals about 20 μ broad, not constricted at the cross-walls, ends slightly attenuated and subcapitata, 3-5 μ broad, spirals about 20 μ distant, cells short, 1/3 as long as broad, cross walls granulated; cells contents with large gas-vacuoles. In cultures.

Genus - **Spirulina** Turpin em. Gardner

Trichomes unicellular or multicellular cylindrical, sheath absent; loosely or tightly coiled into a more or less regular spiral; apex of trichome usually not attenuated; cross-walls if present obscured; terminal cell rounded, without calyptra.

71. Spirulina gigantea Schmidle

Plate-VI, Fig.71

Trichome 3 - 4 µ broad, deep blue-green, regularly spirally coiled, at the end comical attenuated, spirals 11 - 16 µ broad.

In water bodies of paddy fields.

72. Spirulina laxissima West, G.S.

Plate-VI, Fig.72

Trichome 0.7 - 0.8 µ broad, blue-green, spirals very loose, but regular, 4.5 - 5.3 µ broad; 17-22 µ distant from each other, end-cells rounded, obtuse.

In cultures.

73. Spirulina major Kutz. ex Gomont

Plate-VI, Fig.73

Trichome 1.2 - 1.7 (-2) µ broad, regularly spirally coiled, blue-green, spirals 2.5 - 4 µ broad and 2.7-5 distant.

In cultures.

74. Spirulina subtilissima Kutz. ex Gomont

Plate-VII, Fig.74

Trichome 0.6 - 0.9 µ broad, regularly spirally coiled, bright blue-green -or yellowish, spirals 1.5 - 2.5 (-2.8) µ broad, distance between the spirals 1.25 - 2 u.

In water bodies of paddy fields.

Genus - **Oscillatoria** Vaucher

Trichome single or forming a flat or spongy free swimming thallus, sheath absent, rarely with a more or less very delicate sheath, motile, mostly by a creeping movement causing rotation on the longitudinal axis; end of trichome distinctly marked, pointed, bent like a sickle or coiled more or less like a screw. Hormogones formed by the division of the trichome.

On moist soils.

75. Oscillatoria anguina (Bory) Gomont

Plate-VII, Fig.75

Thallus dark blue-green; trichome straight, at the ends spirally coiled and distinctly attenuated, not constricted at the cross-walls, 6 - 8 µ broad, cross-walls sometimes granulated; cells 1/3 - 1/6 as long as broad, 1.5 - 2.5 µ long; end-cells capitate, with a slightly thickened membrane.

On moist soil.

76. Oscillatoria amoena (Kutz.) Gomont

Plate-VII, Fig.76

Thallus more or less blue-green; trichomes straight, slightly constricted at the cross-walls, ends gradually attenuated, 2.5 - 5 µ broad, dull blue-green; cells nearly as long as broad, 2.5 - 4.2 µ long, septa granulated, end cells capitate, broadly conical with calyptra.

In cultures.

77. Oscillatoria animalis Ag. ex Gomont

Plate-VII, Fig.11

Thallus dark blue-green; trichomes straight, not constricted at the cross-walls, briefly attenuated at the ends and slightly bent, 3-4 µ broad, blue-green, cells mostly shorter upto ½ as long as broad, seldom longer, 1.6 - 5 µ long, not granulated at the

cross-walls; end cell pointed conical or acute, no calyptra, not capitate.
In the collection of green mats from soil surface.

78. Oscillatoria acuta Bruhl et Biswas, orth. mute Geitler

79. Plate-VII, Fig.77

Trichomes either solitary or a number of them parallel to each other aggregated into bundles of moderate size, hardy, brittle, not constricted at the cross-walls, 4-6 μ thick, 70-400 μ long, usually quite straight, narrow or accuminate towards the subobtuse, non-capitate, non-calyptrate apex, which may be straight but is more often rather abruptly bent aside; cells 3 - 4 μ long, contents bluish green, finely granular, sometimes with some larger granules close to the surface.
In cultures.

80. Oscillatoria annae van Goor

Plate-VII, Fig.79
Trichome straight, dull blue-green, slightly constricted at the cross-walls, 7.5 - 8 μ broad; mostly attenuated at the ends, up to 7 μ broad and bent; calls 1/3 - 1/5 as long as broad, 1.5 - 3 sometimes upto 4 μ long, not granulated at the cross walls; end-cell rounded, calyptra absent.
On moist soil.

81. Oscillatoria amphibia Ag. ex Gomont.

Plate-VII, Fig.80
Thallus deep blue green; trichome straight or coiled, apices not attenuated, not capitate, not constricted at the cross-walls, 2-3 (3.5) μ broad, pale blue-green; cells 2-3 times longer than borad, 4-8.5 μ long, with two granules at the septa, end cell not capitate, rounded, calyptra absent.
On the banks of water bodies of fields.

82. Oscillatoria amphigranulata van Goor

Plate-VII, Fig.81

Trichome straight, distinctly constricted at the cross-walls, 1.75-2 μ broad, ends not attenuated, not capitate; cells 2.5 - 5 μ long, as long as broad or up to twice as long, with two gas-vacuoles at the septa, pale blue-green; end cell rounded, calyptra absent.

In blue green mats on moist soil surface.

83. Oscillatoria brevis (Kutz.) Gomont.

Plate-VII, Fig.82

Thallus expanded, olivaceous; trichome blue-green, straight, not constricted at the cross-walls, ends briefly attenuated, more or less bent, not capitate, 4-6.5 μ broad, blue-green; cells 1/2 -1/3 times as long as broad, 1.5 - 3 μ long, not granulated at the septa; end-cell rounded, conical, calyptra absent.

In cultures.

84. Oscillatoria biswasii Kamat

Plate-VII, Fig.83

Plant mass blue green; trichomes bent, tapering towards the apices, constricted at the cross walls; cells shorter than broad, or as long as broad, 7-9 μ in diam., 4-7 μ long; end cell conical without cap or calyptra.

In collection of algal mats from paddy fields.

85. Oscillatoria bharadwajae Kamat.

Plate-VII, Fig.84

Plant mass blue green; trichomes long, slightly tapering at the ends, not constricted at the cross walls; cells much shorter than broad, 34 - 38 μ in diam., 3 - 5 μ long; end cell rounded with calyptra.

In cultures.

86. Oscillatoria curviceps Ag. ex Gomont

Plate-VII, Fig.85

Thallus light or dark blue-green; trichomes more or less straight, bent at the end or spirally coiled, not attenuated or very little attenuated, not constricted at the cross-walls, 10 - 17 μ broad, cells 1/3-1/6 as long as broad, 2-5 μ long, cross-walls sometimes granulated; end-cells flat rounded, not capitate.

In cultures.

87. Oscillatoria curviceps Var. **angusta** Ghose

Plate-VII, Fig.86

Trichome 5-10 μ broad.

In cultures.

88. Oscillatoria carakiana playfair. V. **non-grnanulata** Kamat

Plate- V I I , Fig. 87

Plant mass blue-green; trichomes long curved, not tapering towards the apices, not constricted, not granulated at the cross-walls; cells much shorter than broad, 8.4 -105 μ in diam., 1.8-3 (-3.5) μ long, sometimes with gas-vacuoles: end cell rounded without cap or calyptra.

In cultures.

89. Oscillatoria cortiana Meneghini ex Gomont

Plate-VII, Fig.88

Thallus dull blue-green; trichome straight, slightly constricted at the joints, 5.5 - 8 μ broad, gradually tapering at the ends, bent, not capitate, blue-green; cells as long as broad or longer or shorter than broad, 5.4 - 8.2 μ long, at the ends up to 14 μ long, septa not granulated; end cell obtuse, without calyptra.

Perceived aligned on plant surface near water levels.

90. Oscillatoria chalybea (mertens) Gomont var. **insularis** Gardner

Plate-VII, Fig.89

Trichomes ends bent and sickle-shaped, 6.4 - 7.2 µ broad, as long as or shorter than broad, not constricted at the cross-walls.

In cultures.

91. Oscillatoria chalybea (Mertens) Gomont. v. **Minor** Kamat

Plate-VIII, Fig.90

Plant mass olive green or violet; trichomes usually straight, slightly tapering at the apices, slightly constricted at the cross walls; cells shorter than broad, 5-5.7 µ in diam., 1.8 - 3.2 µ long; end cell peak-like, without cap or calyptra.

In cultures.

92. Oscillatoria chilkensis Biswas

Plate-VIII, Fig.91

Trichomes 4 µ diam., somewhat curved, slightly constricted at the joints, apex of the trichomes very shortly tapering, very rarely undulated, not hooked; apical cell obtusely rounded, not pointed or capitate, calyptra none; cells shorter than the diameter, about 2 µ in length, transverse walls not granulated, cell-contents homogeneous, pale blue-green.

In cultures.

93. Oscillatoria chlorina Kutz. ex Gomont

Plate-VIII, Fig.92

Thallus very thin, yellowish green; trichome straight or curved, unconstricted or slightly constricted at the cross- walls; 3.5-4 µ broad, sometimes up to 6 µ broad, gas-vacuoles absent; cells somewhat longer or shorter than broad, 3.7 - S μ long, cross-walls not granulated; calyptra absent. In algal mats collections from moist soil of paddy field.

94. Oscillatoria decolorata West, G.S.

Plate-VIII, Fig.93

Trichome single, nearly straight, short indistinctly coloured, 12 - 13 μ broad, not attenuated, not capitate, end cell hemispherical, cells somewhat shorter than broad, not constricted at the cross-walls, with reticulate contents.

In cultures.

95. Oscillatoria foreaui Fremy ,

Plate-VIII, Fig.94

Trichomes sparse, elongate, bent or suberect, apex gently curved, subfalsiform, distinctly constricted at, the cross-walls and more or less torulose, 3 μ broad, apex not attenuated not capitate; cells 1/2 as long as broad, protoplasm granular, septa commonly inconspicuous, apical cell obtuse conical, calyptra absent.

In cultures.

96. Oscillatoria geitleriana Elenkin

Plate- VIII , Fig. 95

Trichome single, amidst other algae, long, flexible, blue-green, 2.5 μ broad, not constricted at the cross-walls, ends erect, not attenuated, subcapitate, cells 1½ times longer than broad, 3 - 3.5 μ long, septa not granulated; and cells with a conves distinctly thick membrane.

In cultures.

97. Oscillatoria grunowiana Gomont

Plate-VIII, Fig.96

Thallus dull green; trichome mostly very lightly spirally coiled, but also straight or irregularly bent, 3.5-6 µ broad, apices not attenuated, or slightly attenuated, slightly capitate; cells 1/3 as long as broad, or nearly as long, 1.4 - 4 µ long, often granulated at the cross-walls; end cell rounded without a calyptra.

In cultures.

98. Oscillatoria irrigua (Kutz.) Gomont.

Plate - VIII, Fig.97.

Thallus blackish blue green, trichome light bluish purple, when dried hyaline or pale blue, straight, flexuous, not torulose, 6-11 µ broad, apex, slightly attenuated subcapitate, straight; cells quadrate to ½ as long as broad, 4-11 µ long, contents in dried specimen, pellucid, granular; septa ordinarily not granulated; apical cell convex, with an evident thickened outer wall.

In cultures.

99. Oscillatoria jasorvensis Vouk.

Plate - VIII, Fig.98.

Thallus pale blue-green, yellowish green, straight, 2.5-3 µ broad, bent at the ends, not attenuated, not capitate; cells as long as broad or nearly so; end cells rounded, without calyptra.

In cultures.

100. Oscillatoria jasorvensis G. Schmid.

Plate - VIII, Fig.99.

Thallus blue-green, trichomes long, usually curved, not constricted at the cross walls, not or very slightly attenuated at the ends; cells shorter than broad, 1/5 - 1/6 times as long as broad, 20-33 µ in diam., 4-6 µ long, end cell round, without a cap or calyptra.

In cultures.

101. Oscillatoria limosa Ag. ex Gomont.

Plate-VIII, Fig.100.

Thallus dark blue-green to brown; trichome more or less straight, dull blue-green, brown or olive-green, not constricted at the cross-walls. or only slightly constricted, 11-20 (-22) µ, commonly 13-16 µ broad; cells 1/3-1/6 as long as broad, 2-5 µ long, cross-walls frequently granulated; end-cell flatly rounded with slightly thickened membrane.

In collections of algal mats from soil surfaces.

102. Oscillatoria limosa Ag. ex Gomont yare **disperso-granulata** Schkorb.

Plate - VIII, Fig.101.

Thallus blue-green, trichomes long, curved, not constricted at the cross walls, slightly attenuated at the ends, cells shorter than broad, upto 1/3-1/4 times as long as broad, 10-17 µ broad, 2.8-4.0 µ long, end cell rounded, without a cap or calyptra.

In cultures.

103. Oscillatoria limnetica Lemm.

Plate-VIII, Fig.l02

Trichome straight or slightly bent, distinctly constricted at the cross-walls, pale blue-green, 1.5 µ broad, not attenuated, not capitate; cells 2 ½ - 6 times as long as broad, 4 - 12 µ long; end cells rounded, calyptra absent.

In cultures.

104. Oscillatoria laete-virens (Crouan) Gomont

Plate-VIII, Fig.l03

Thallus thin, membranous, green; trichome yellowish green, straight, fragile, slightly constricted at the cross- walls 3 - 5 µ broad, apices attenuated, undulate or bent; cells nearly as long

as broad, 2.5 - ~5 μ long, sometimes granulated at the cross-walls; end-cells not capitate, more or less obtuse or conical, without calyptra.
Perceived near the banks of water bodies in paddy field.

105. Oscillatoria laete-virens (Crouan) Gomont Var. **minimus** Biswas

Plate-VIII, Fig.l04
Trichomes 2.5 - 3 μ in diam., somewhat fragile, slightly constricted at the cross-walls, apex of the trichome slightly tapering, more or less curved, and sinuous, not distinctly hooked; apical cell acute and somewhat pointed, not capitate, calyptra none; cells 1.5 - 2 μ in length; cross-walls granulated, 3 granules on either side, cell con-tents uniformly granular, blue-green.
In cultures.

106. Oscillatoria mouqeotii Kutz.

Plate-VIII, Fig.105
Thallus dull blue-green, at first attached, later free-swimming; trichomes straight, or somewhat bent, constricted at the cross-walls, 5.5 - 7.5 μ broad, not attenuated, end cell broadly rounded.
In cultures.

Oscillatoria nigroviridis Thwaites ex Gomont

Plate - V III, Fig. 10 6
Thallus expanding, dark green. Trichome olive-green, 7 - 11 μ broad, constricted at the cross-walls, bent at the ends and attenuated, cells 1/2 - 1/4 as long as broad, 3 - 5 μ long, cross-walls granulated, end-cells slightly capitate with a slightly thickened convex outer membrane.
In cultures.

107. Oscillatoria nigra Vaucher

Plate-VIII, Fig.107

Thallus more or less leathery, often free-swimming, olive or dark-brown or blackish, glistening; trichomes straight or more or less bent, 8.5 µ (6-9 (-10) µ broad), ends round and truncated, cells as long as broad or up to 1/3 as long as broad, septa granulated, slightly constricted.

In cultures.

108. Oscillatoria obscura Bruhl et Biswas

Plate-VIII, Fig.l08

Trichome about 4 µ broad, attenuated at the apex, rounded, slightly bent or nearly straight, blue-green, not constricted at the cross-walls; cells about 1/5 as long as broad or shorter, cross-walls granulated.

In cultures.

109. Oscillatoria okeni Ag. ex Gomont

Plate-IX, Fig.109

Thallus dull blue-green; trichome straight, fragil distinctly constricted at the cross-walls, 5.5 - 9 µ broad, at the ends gradually attenuated, undulating, slightly bent; cells 1/3 as long as broad, 2.7-4.5 µ long, at the ends up to 8 µ long; end cells obtuse or subconical not capitate, without calyptra.

On the moist soil surface of paddy fields.

110. Oscillatoria ornata Kutz. ex Gomont

Plate-IX, Fig.110

Thallus dark blue-green; trichome spirally coiled at e the ends, constricted at the cross-walls, 9-11 µ broad, dull blue-green, cells 1/2-1/6 as long as broad, 2-5 µ long, cross-walls

granulated; apices slightly attenuated; end-cells rounded, not capitate, without thickened membrane.
In collections of algal mats from wet paddy field.

111. Oscillatoria princeps Vaucher ex Gomont

Plate-IX, Fig.lll
Trichomes blue-green, more or less brownish, violet or reddish, mostly forming a thallus, mostly straight, not constricted at the cross-walls, 16 - 60 μ broad, commonly 25 - 50 μ, blue-green to dirty green, slightly or briefly attenuated at the apices and bent; cells 1/11-1/4 as long as broad, 3.5 - 7 μ long; and-cells flatly rounded, slightly capitate without or with slightly thickened membrane.
In collections as well as cultures.

112. Oscillatoria proboscidea Gomont

Plate-IX, Fig. 112,
Thallus dull green to dark blue-green; trichome more or less straight, not constricted at the cross-walls, 12-15 μ broad, at the ends distinctly attenuated, slightly curved or some tunes spirally coiled, brightly blue-green; cells 1/3 1/6 times as long as broad, 2-4 μ long, not granualated at the cross-walls; end-cells flatly rounded, capitate, with slightly thickened membrane.
In cultures.

113. Oscillatoria profunda Kirchner.

Plate - IX, Fig.113.
Thallus dark blue-green; trichomes long, usually straight, sometimes bent only at the apices, not constricted at the cross walls, not attenuated at the ends; cells as long as or longer than broad, 1.5-2.1 μ broad, 2.1-3.5 (-5.2) μ long end cell rounded without a cap or a calyptra.
In cultures.

114. Oscillatoria prolifica (Grev.) Gomont.

Plate - IX, Fig.114.

Trichomes blue-green, straight, 3.5-5.5 μ broad; cells 2.5-4.0 μ long, septa granulated, end cell capitate with a calyptra.

In cultures.

115. Oscillatoria pseudogeminata G. Schmid.

Plate - IX, Fig.115.

Thallus pale or dirty blue-green, trichomes coiled, pale blue-green, ends not attenuated, 1.3 - 2.2 μ broad; cells as long as broad or somewhat longer or shorter than broad, about 2.6 μ long, not constricted at the cross-walls, cross-wall thick, not granulated, end cell rounded, calyptra absent.

In cultures.

116. Oscillatoria quadriaunctulata Bruhl et Biswas

Plate-IX, Fig.116

Trichomes curved or nearly straight, closely associated into a delicate very pale greenish blue membrane, 1 - 1.5 μ in diam., cells 3.5 - 5 μ long, 2-4 times as long as wide, not constricted at the joints, contents minutely granular, transverse walls rather obscure marked by a pair of somewhat larger granules on either side.

In collections as well as in cultures.

117. Oscillatoria rubescens DC ex Gomont

Plate-IX, Fig.117

Trichome straight, at the ends gradually attenuated, 6 - 8 μ broad, not constricted at the cross-walls, sometimes forming a purple red, or violet, free-swimming bundles; cells 1/2 - 1/3 as long as broad, 2 - 4 μ long, often granulated at the septa, with gas-vacuoles; end cell capitate, with convex calyptra.

In collections.

118. Oscillatoria rubescens DC ex Gomont. Var. **Curvata** Kamat

Plate-IX, Fig.118

Plant mass blue green; trichomes long, attenuated towards the apices, not constricted at the cross walls; cells much shorter than broad, 1/5 - 1/3 times as long as broad, 5 - 7.2 µ in diam., 1.4 - 2.1 µ long; end cell capitate with calyptra.

In cultures.

119. Oscillatoria raoi De Toni, J.

Plate-IX, Fig.119

Plant mass thin, membranous, firm, pale blue-green to pale bluish violet; trichome straight, usually of uniform thickness, and only rarely slightly tapering at the ends, without constrictions at the joints, 5.2 - 6 µ broad, septa indistinct, but with distinct granules closely arranged on either side; cells 2.5-6 µ (average 5 u) long, with homogeneous contents, end cells rounded rarely conical, sometimes with constrictions at the septum, not capitate, without any calyptra.

In cultures.

120. Oscillatoria raciborskii Wolosz.

Plate-IX, Fig.120

Trichome single, yellow-green, very long, not constricted at the cross-walls, 5-9 µ broad, attenuated at the ends, curved; cells ½ times as long as broad, at the ends, filaments nearly quadrate, with gas-vacuoles; and cell painted conical, without calyptra.

In cultures.

121. Oscillatoria schultzii Lemm.

Plate-IX, fig.121

Thallus dull-green to very dark; trichomes straight or bent, distinctly constricted at the crass-walls gradually attenuated at

the apices, straight or bent like a sickle, 2.6 µ broad, pale blue-green; cells nearly barrel-shaped as long as broad or somewhat longer seldom somewhat shorter than broad, 1.5-4 µ long, granulated at the crass-walls; end cell conical.
In cultures.

122. Oscillatoria Schulzii Lemm. V. **cylindrica** Kamat

Plate-IX, fig.122
Plant mass dark green, trichomes curved, tapering towards the apices, constricted at the crass-walls; cells as long as broad or longer than broad, (1.5-) 2-3 µ in diam., 2.1-3.5 µ long; end cell 4.2-5 µ long, conical without cap or calyptra.
In cultures.

123. Oscillatoria subuliformis Kutz. ex Gomont

Plate-IX, Fig.123
thallus dull green; trichomes yellow-green, very long flexuous, and bent, not constricted at the cross-wall, 4.7-6.5 µ broad, at the ends gradually attenuated and bent; cells nearly quadrate, 4.7-6.5 µ long, at the ends up to 10 µ long, obtuse, not capitate, calyptra absent.
In cultures.

124. Oscillatoria subbrevis Schmidle

Plate-IX, Fig.124
Trichomes single, 5-6 µ broad, nearly straight, not attenuated at the apices; cells 1 - 2 µ long, not granulated at the cross-walls; end-cell rounded, calyptra absent.
In collections of algal mats from paddy s fields.

125. Oscillatoria subbrevis Schmidle f. **crassa** Dixit

Plate-IX, Fig.125
Trichomes, 12.5 - 20 µ broad, cells 2.5 - 3 µ long.
In cultures.

126. Oscillatoria subbrevis Schmidle f. **major** West, Gomont

Plate-IX, Fig.126
Trichome 10 - 10.5 μ broad.
In cultures.

127. Oscillatoria subtilissima Kutz.

Plate-IX, Fig.127
Trichome single or a few together, seldom forming a thallus, yellowish green, 1-1.5 μ broad, straight or curved, septa indistinct, without gas-vacuoles.
In cultures.

128. Oscillatoria splendida Grev. ex Gomont

Plate-X, Fig.128
Thallus brilliant blue-green or olive-green; trichome straight or curved, not constricted at the cross- walls, at the ends gradually attenuated, 2 - 3 μ broad, cells 2-4 times longer than broad rarely quadrate, 3-9 μ long, septa often granulated; ends more or less bent, sometimes screw-like or a sickle; end cells capitate, nearly rounded, mostly without calyptra.
In cultures.

129. Oscillatoria splendida Grev. ex Gomont Var. **attenuata** W. and G.S. West

Plate-X, Fig.129
Trichome 2 - 2.3 μ broad, longer than the type, ends prominently attenuated.
In cultures.

130. Oscillatoria salina Biswas

Plate-X, Fig.130
Plant mass forming a deep blue-green thin membrane extending over the muddy soil and finally after being separated floating

on the surface of the water; filaments lying side by side in the stratum, straight, elongate, erect, scarcely curved, fragile, rapidly moving, not at all constricted at the joints, 3 - 5 µ dia., apices of trichome straight, briefly tapering ending acuminately in a sharp point, hooked or twisted, not capitate; apical cell mucronate hyaline, calyptra absent; cells shorter than broad, l.5 - 2 µ long, sometimes filament may be interrupted by inflated refringement cells; transverse septa indistinct, not granulated, cell contents finely uniformly granular, almost homogeneous, blue-green.

In collections as well as in cultures.

131. Oscillatoria sancta (Kutz.) Gomont

Plate-X, Fig.131

Thallus dark blue, shining, thin, gelatinous; trichomes straight or bent, distinctly constricted at the cross-walls; ends briefly attenuated, 10 - 20 µ broad, dull blue-green or olive-green; cells 1/3 - 1/6 times as long as broad, 2.5 - 6 µ long, granulated at the cross-walls. End-cell flattened, hemispherical, slightly capitate, with a thickened membrane.

In cultures.

132. Oscillatoria terebriformis Ag. ex Gomont

Plate-X, Fig.132

Thallus dull blue; trichomes end bent in a screw like manner, and slightly attenuated, unconstricted at the cross-walls 4 - 6.5 µ broad, 2.5 - 6 µ long; end cell rounded, not capitate, calyptra absent.

In cultures.

133. Oscillatoria tenuis Ag. ex Gomont

Plate-X, Fig.133

Thallus thin blue-green or olive-green, slimy; trichome straight, fragile slightly constricted at the cross- walls, 4-10 μ broad, blue-green, sometimes bent at the ends, not attenuated at the apices, not capitate; cells up to 1/3 as long as broad, 2.6 - 5 μ long, at the septa mostly granulated; end cell more or less hemispherical with thickened outer membrane.
In cultures.

134. Oscillatoria vizagapatensis Rao, C.B.

Plate-X, Fig.134
Thallus blue-green; trichomes straight, or bent, pale blue-green, uniformly broad except at the extreme apex, 8-10 μ broad, without constrictions at the cross-walls; cells much shorter than broad, 1.6-2 μ long, contents granular; end cell broadly rounded forming a cap with a slightly thickened outer wall.
In cultures.

Genus - **Phormidium** Kutz

Filaments many forming a gelatinous or leathery stratum, thallus attached by the lower side, or floating in lJater with torn margins; sheath present, more or less firm, sometimes agglutinated, sometimes partly diffluent, thin, colourless; trichomes cylindrical, in some constricted at the joints, apices often attenuated, straight or bent, never regularly spirally coiled, capitate or non-capitate, apical cells in many species with a calyptra.

135. Phormidium africanum Lemm.

Plate-X, Fig.135
Thallus thin, blue-green; filaments straight or variously bent; sheath diffluent; not coloured violet by chlor-zinc-iodide; trichome pale blue-green, distinctly constricted at the cross-

walls, not attenuated at the ends, 1.5-2 μ broad; cells quadrate or rectangular, 1.5-4 μ long; end cells truncated, with a small calyptra.
In cultures.

136. Phormidium angustissimum W. et G.S. West

Plate-X, Fig.136
Thallus leathery, thin, pale blue-green; trichomes bent, entangled, constricted at the cross-walls, ends not attenuated, straight, 0.6-0.8 μ broad, pale blue-green; sheath colourless, agglutinated; cells cylindrical, 2 – 8 mostly 4 - 5 times as long as broad, septa not granulated, apical cell not capitate.
Perceived aligned to stalks of paddy below water.

137. Phormidium anomala Rao, C.B.

Plate-X, Fig.137
Thallus thick expanded, soft, mucilaginous, deep blue-green, 3 - 6 mm thick; trichome subparallel of uniform width, 8 - 10 μ (-10.5) μ broad without constrictions at the cross-walls; sheath thin, colourless, not stained by chlor-zinc-iodide, persistant or dissolved; cells disc-shaped, much broader than long, 0.8 - 1.2 (-2) μ long; end cells bluntly rounded without cap or calyptra.
In cultures.

138. Phormidium ambiquum Gomont

Plate-X, Fig.138
Thallus more or less expanded, bright blue-green dark or yellowish green; filaments flexuous, variously entangled; trichomes slightly constricted at the cross-walls. at the ends not attenuated, not capitate, 4 - 6 μ broad, blue-green; sheath thin, firm or diffluent, sometimes thick and more or less lamellated, cells shorter than broad, 1.5-2.7 μ long, rarely granulated at

the cross-walls, sometimes with gas-vacuoles; end cell rounded, calyptra absent.

Perceived aligned to fallen leaves of paddy below water.

139. Phormidium abronema Skuja

Thallus coriaceous, lamellated, rarely mucosa flocculose, olivaceous, blackish green to light bluish; filaments flexuous, more or less loosely spirally coiled, 4-7 μ broad, sheath moderately thick, hyaline, mucilaginous, commonly diffluent; trichome 3-4.5 μ broad, ends briefly attenuated, distinctly constricted at the cross-walls, cells cylindrical, barrel-shaped, 3/4-2 times as long as broad, septa not granulated, end cell more or less rounded conical.

In algal mat collections from wet soils.

140. Phormidium bohneri Schmidle

Plate-X, Fig.140

Thallus thin, mucilaginous, green; trichome nearly straight, more or less parallel, not constricted at the cross- walls, 1.7 - 2 μ broad, not attenuated at the ends, not capitate, blue-green; sheath hyaline, diffluent, cells mostly shorter than broad, seldom quadrate or somewhat longer than broad; end-cell rounded.

On moist soil surfaces.

141. Phormidium ceylanicum Wille

Plate-X, Fig.141

Thallus olive blue-green, or dull green, flat; sheath diffluent, in a homogeneous mucilage, not coloured violet by chlor-zinc-iodide; trichome blue-green, straight, slightly constricted at the cross-walls, ends gradually tapering, not capitage, 7-9 μ broad; cells as long as broad or 1/2 - 1/3 times as long; end cell more or less pointed by conical, calyptra absent.

In cultures.

142. Phormidium ceylanicum Wille. Var. **minor** Kamat

Plate-X, Fig.142

Plant mass blue green; sheath diffluent; trichomes attenuated, towards the apices, constricted at the cross- walls; cells shorter than broad or as long as broad, 3 - 4 µ in diam., 3.5 - 4 µ long; end cell conical, beak like without calyptra.

In cultures.

143. Phormidium corium (ag.) Gomont. Var. **capitatum** Gardner

Plate-X, Fig.143

Sheath thin, firm and smooth; trichomes 5.2 - 6.4 µ broad, not constricted at the cross-walls, pale blue-green; end cell round or obtuse conical, with distinctly thickened outer membrane.

In cultures.

144. Phormidium cebennense Gomont

Plate-X, Fig.144

Thallus thin, expanded, golden brown; filaments flexuous, entangled or nearly parallel and straight; sheath thin, papyraceous, firm or diffluent, not coloured violet by chlor-zinc iodide; trichome light brown, 2 µ broad, not constricted at the cross-walls, ends not attenuated, not curved; cells 1-1.5 µ long, septa not granulated; end-cell rounded, calyptra absent.

In cultures.

145. Phormidium dimorphum Lemm.

Plate-XI, Fig.145

Thallus membranous, broadly expanded, dark blue-green; trichome straight or bent, sometimes arranged parallel, distinctly constricted at the cross-walls, 3.5 - 5 µ broad, blue-green, at the ends long and prominently attenuated; sheath firm, colourless, thin, not lamellated, close, cells mostly quadrate, seldom shorter, also longer than broad at the ends, 2-5.5 mostly 3 μ long, not granulated at the cross-walls; end cells long, conical or short rounded and hemispherical but truncated about 2 μ broad and 1.5 μ high calyptra.

On moist soil surface of paddy fields.

146. Phormidium foveolarum (Mont.) Gomont

Plate-XI, Fig.146

Thallus thin, dark-green; trichome flexuous, constricted at the cross-walls, at the ends not attenuated, about 1.5 μ broad, pale blue-green; sheath colourless, diffluent in an amorphous gelatinous mucilage, cells nearly quadrate or somewhat shorter than broad, 0.8-1.8 μ long, septa not granulated; end cell rounded, calyptra absent.

On moist soil surface as well as in soil cultures.

147. Phormidium fragile (Meneghini) Gomont

Plate-XI, Fig.147

Thallus mucilaginous, lamellated, yellowish or brownish blue-green; sheath diffluent, trichomes more or less flexuous, entangled or nearly parallel, 'diestinctly constricted at the cross-walls, septa not granulated, attenuated at the ends, 1.2-2.3 μ broad, pale blue-green; cells nearly quadrate, 1.2-3 μ long; end cell acute-conical, calyptra absent.

In algal mat collected from wet soils of paddy fields.

148. Phormidium favosum (Bory) Gomont

Plate-XI, Fig.148

Thallus dark blue-green, when dried dark steelblue coloured, generally a little expanded, papery attached to the base; trichomes mostly without sheath in an amorphous mucilage, blue-green, elongate, more or less flexuous, not constricted at

the cross- walls; (4-) 4.5 - 9 μ broad, ends straight or loosely spirally coiled, gradually attenuated, capitate; cells quadrate to up to 1/2 as long as broad, 3-7 μ long; cross-walls granulated, end cell obtuse truncated, calyptra subhemispherical.
In cultures.

149. Phormidium hieronymusii lemmermann

Plate-XI, Fig.149
Thallus dull olive-green; trichome more or less regularly spirally coiled, not constricted at the cross-walls, distinctly attenuated at the ends, 7 μ broad, dull blue or yellow-green; sheath thin, firm, cells very short, 1.5 - 2 μ long, cross-walls often granulated; end cell rounded, seldom slightly capitate.
In cultures.

150. Phormidium jenkelianum Schmid, G.

Plate-XI, Fig.150
Thallus brownish black, slimy; trichome brownish blue-green, flexuous, 2 - 2.6 μ broad, not attenuated at the ends; cells 1/2 - 1/3 as long as broad, 0.6-1.3 μ long, well constricted at the cross-walls, septa not granulated, sheath delicate, end cell rounded, truncated, calyptra absent.
In cultures as well as collections of algal mats.

151. Phormidium luridum (Kutz.) Gomont

Plate-XI, Fig.151
Thallus membranous, on the surface purple to dark violet to blackish to blue-green and grey to blue-green on the under surface; trichome flexuous, fragile, ends not curved, not attenuated, slightly constricted at the cross-walls, 1.7-2 μ broad, light violet, septa not granulated; cells 1.8- 4.7 μ long; sheath thin, diffluent, but later becoming compact, end-cell rounded, calyptra absent.

In cultures.

152. Phormidium mucosum Gardner

Plate-XI, Fig.152

Filaments 7.2 - 7.8 μ broad, long, straight, or curved; trichome 2.5 - 3.4 μ broad, not constricted at the cross-walls, ends; sheath apparently thick, more or less gelatinous, colourless, unlamellated; cells 2 - 2.1/3 times as long as broad, pale blue-green; end cell rounded.
In cultures.

153. Phormidium mucosum Gardner. Var. **arvense** Rao, C.B.

Plate-XI, Fig.153

Plant mass thick, fragile, blue-green to grayish blue; filaments loosely entangled, 3.2 - 5.2 μ broad; sheath thick, firm, hyaline, unstratified, up to 2 μ thick; trichomes of uniform width, not constricted at the joints, 1.8 - 2.5 μ broad, apical cells with rounded apices, without calyptra or cap cells; cells quadratic or almost quadratic or slightly longer than broad, 1.5 - 3.6 μ long.
In cultures.

154. Phormidium molle (Kutz.) Gomont

Plate-XI, Fig.154

Thallus mucilaginous, thin, light blue-green, sheath more or less diffluent, colourless, trichome straight or nearly straight, distinctly constricted at the cross-walls, not attenuated at the ends, 2.7 - 3.3 μ broad, light blue-green, septa not granulated; cells quadrate, cylindrical or pressed barrel-shaped shorter or longer than broad, 3-7.8 μ long, end cell rounded, calyptra absent.
Perceived on the edges of water bodies in paddy fields.

155. Phormidium molle (Kutz.) Gomont f. **tenuior** W. & G.S.West

Plate-XI, Fig.155
Cells 2 - 2.4 μ broad, 2.2 - 3.5 μ long.
In cultures.

156. Phormidium mucicola Hub. - Pestalozzi et Naumann

Plate-XI, Fig.156
Filaments short, straight, 10-20 μ occasionally up to 50 μ long; sheath very thin, trichome not attenuated at the ends, more or less distinctly constricted at the cross- walls; cells 1.5 - 2 μ seldom 1.3 μ broad and more or less twice as long as broad; end cells rounded seldom slightly conical, calyptra absent; contents granulated, pale bluish.
In the green water collected from water bodies of paddy fields.

157. Phormidium orientale West, G.S.

Plate-XI, Fig.157
Thallus very thin, blue-greenl filaments more or less entangled; sheath very thin, firm; trichome 1.6 - 1.8 μ broad, not constricted at the cross-walls; ends straight, not capitate; cells 3 ½ - 4 ½ times as long as broad.
In cultures.

158. Phormidium papyraceum (Ag.) Gomont

Plate-XI, Fig.158
Thallus dark green, glistening, expanding, thin, leathery, when dry fragile, filaments elongate, very flexuous, densely entangled; sheath thin; papery, persistent sometimes diffluent, not constricted at the cross-walls, 3 - 5 μ broad; ends straight, briefly attenuated not capitate; cells sub- quadrate or mostly

shorter than broad, 2 - 4 μ long; cross- walls conspicuous, not granulated, end cell obtuse conical, calyptra absent.
In cultures.

159. Phormidium stagnina Rao, C.B.

Plate-XI, Fig.159
Thallus soft blue-green and membranous; filaments 12.8-14.4 μ broad; trichome blue-green, interwoven and not attenuating, 8 - 9.6 μ broad; sheath hyaline, thick, firm, sometimes diffluent; cells small, without constrictions at the joints, 1.3 - 2 μ long; end cell broadly rounded with a prominent calyptra.
Perceived floating algal mass in water bodies of algal mat.

160. Phormidium tenue (Mensgh.) Gomont

Plate-XI, Fig.160
Thallus pale blue-green, thin, membranous, expanded; trichome straight or slightly bent, densely entangled, slightly constricted at the cross-walls, attenuated at the ends, 1-2 μ broad, pale blue-green; sheath thin, diffluent, cell up to 3 times longer than broad, 2.5-5 μ long, septa not granulated, cross-walls not commonly visible; end-cell acute-conical, calyptra absent.
In cultures.

161. Phormidium uncinatum (Ag.) Gomont

Plate-XI, Fig.161
Stratum broadly expanded, dark green to brownish black, adherent, thin, firm, or floating attached at the base, thick, torn; filaments straight or slightly bent, sheath mucilaginous, distinct or diffluent in an amorphous mucilage, trichome blue-green, not constricted at the cross-walls, 6 - 9 μ broad, ends briefly attenuated, capitate, curved or short spirally coiled; cells 1/2 - 1/3 as long as broad, rarely subquadrate, 2 - 6 μ long,

cross-walls frequently granulated; end-cell with a round or depressed conical calyptra.
In cultures.

Genus - **Lyngbya** Ag.

Trichome single or free in a thin or very massive thick, firm sheath; sheath mostly colourless, seldom coloured yellow to brown or red, blue to purple red; filaments sometimes spirally coiled or attached at the base or in the middle or the entire filament attached, mostly without such attachment or free-swimming or forming free thallus.

162. Lyngbya attenuata Fritsch.

Plate - XI, Fig.162.
Thallus dark blue-green, filaments 5-5.2 μ broad, long, entangled; sheath thin, unstratitied, hyaline to yellow brown.; trichomes not constricted at the cross-walls. attenuated the ends; cells as long as or shorter than broad, 3.8-4.5 μ broad, 3.8-4.5 μ long; end cell conical, without a cap or a calyptra.
In cultures.

163. Lyngbya aestuarii Liebm. ex Gomont Var. **tenuis** Dixit

Plate - XI, Fig.163.
Filaments 9-12.5 μ diam; trichomes 5-9 μ broad; cells 1.25-2.5 μ long, no gas-vacuoles, cross-walls granulated.
In cultures.

164. Lyngbya allorgei Fremy.

Plate - XII, Fig.164.
Filaments solitary or united and caespitose, fasciculate; tortuous, intricate, elongate, sheath very thin, papyraceous, colourless, trichome pale violet, not constricted at the cross-walls, 3.5 - 4 μ broad, cells nearly quadrate or up to 1 ½ times

as long as broad, cross-walls not granulated; end cell rotund, calyptra absent.

In soil cultures.

165. Lyngbya aerugineo-coerulea (Kutz.) Gomont

Plate-XII, Fig.165

Filament's single or more rarely forming a dull blue-green expanded thallus, flexuous, fragile; sheath thin, firm, not lamellated, trichomes 4 - 6 μ broad, not constricted at the cross-walls, sometimes granulated, apex of trichome occasionally capitate; cells 1-1/2 as long as broad, 2.3-3 μ long, pale blue-green; end cell flattened, conical or rotund, with a slightly thickened outer membrane.

In cultures as well as collections from water bodies of paddy field.

166. Lyngbya cryptovaginata Schkorbatow

Plate-XII, Fig.166

Filaments single, free-floating, straight, 4-9 μ broad; sheath colour less delicate, at the first often not distinctly visible but later (in cultures) distinct, trichome slightly constricted at the cross-walls, blue-green; cells nearly quadrate, or up to ½ as long as broad, with pseudovacuoles; end cells rounded.

In collections from water bodies of paddy fields.

167. Lyngbya ceylanica Wille

Plate-XII, Fig.167

Thallus olive-green, violet or red; filaments 10 - 14 μ broad, straight; sheath thin, colourless, when older often red, trichome blue-green or violet, unconstricted at the cross-walls, not attenuated at the ends, 8 - 12 μ broad, cross-walls not granulated; cells quadrate to 1/2 or 1/3 as long as broad; end cell rotund, without calyptra.

Perceived on moist soil surfaces.

168. Lyngbya corticicola Bruhl et Biswas

Plate-XII, Fig.168

Thallus a thin tomentose dark or yellowish brown layer; filaments somewhat fragile, moderately flexuous, more or less densely intricate, 12 - 26 μ thick, sheath at first hyaline, but later becoming brown, 2 μ thick; scarcely or not at all lamellated, surface uneven, not wrinkled; trichomes 8-12 μ broad, slightly constricted at the joints, cells about 1/2-1/3 as long as wide, apex rotund, contents green, granular, dissepiments not granulated. On moist soil as well as aligned to paddy leaves beneath water.

169. Lyngbya contorta Lemm.

Plate-XII, Fig.169

Filaments single, free-floating, regularly spirally coiled, with a delicate, nearly circular coils, 1 - 1.5 μ broad; sheath narrow, colourless; cells 1-2 μ broad, 3 - 5 μ long, not constricted at the cross-walls, granulated with a single granule or without them; end cell rounded, not attenuated.

In cultures.

170. Lyngbya dendrobia Bruhl et Biswas

Plate-XII, Fig.170

Stratum more or less expanded, compact, thin, minutely and densely tomentose; filaments long and flexible; closely interwoven, with sheath 10-11 μ thick; sheath usually thin, 1 - 1.5 μ thick, smooth, hyaline, usually colourless, more rarely when old brownish and very moderately stratified; trichomes 9-10 μ bro§d, scarcely or not at all constricted at the cross-walls; cells 1.7 - 2.5 times as broad as long, 4-6 μ long; contents of various shades of brown, uniformly and densely granular; dissepiments conspicuous, not marked by granules.

In cultures.

171. Lyngbya dendrobia Bruhl et Biswas .f. **lurida** Bruhl et Biswas

Plate-XII, Fig.171
Cells 10 - 12 μ diam., 6 - S μ long; sheath 0.6 μ thick, broad, hyaline.
In cultures.

172. Lyngbya dendrobia Bruhl et Biswas Var. **skujaii** nom. novo

Plate-XII, Fig.172
Filaments variously long, more or less irregularly coiled, sometimes also straight, 6.5 - 7 μ broad; sheath firm and smooth, very thick, about 0.9 μ thick, not gelatinised, sometimes with distinct parallel lamellation, generally colourless, occasionally light to dull ambre brown; trichomes 4 - 6 μ broad, unconstricted to slightly constricted at the cross-walls, not attenuated at the ends; cells mostly as long as broad, also longer or shorter, ¾ - 1½ times the breadth, septa not granulated; end cell rounded and conical, with a somewhat thickened outer membrane.
In cultures.

173. Lyngbya diqueti Gomont

Plate-XII, Fig.173
Filaments forming a caespitose, light green, 2 mm thick thallus, tortuous and entangled in their basal portions, at the ends straight, 2.5 - 3 μ broad; sheath thin, colourless, trichome not constricted at the cross-walls, 2 - 3 μ broad; cells nearly quadrate or seldom shorter than broad, 1 - 3.7 μ long; end cells rotund, without calyptra.
In collections of algal mats from wet soil.

174. Lyngbya gardneri (Setchell et Gardner) Geitler

Plate-XII, Fig.174

Filaments 1 - 2 mm long, attached by a basal cell, forming a thick or dense, straight or slightly bent, 2.6-3 µ broad; trichome 1.3 - 1.6 µ broad, slightly constricted at the cross-walls; sheath indistinct, colourless, smooth; cells quadrate or somewhat longer or shorter than broad; end cells rounded.

In soil cultures.

175. Lyngbya hieronymusii Lemm.

Plate-XII, Fig.175

Filaments single, free-floating, straight or slightly bent, 12 - 14 µ broad; sheath firm, homogeneous, colourless, cells 11 - 13 µ broad, 2.5 - 4 µ long, not constricted at the cross-walls, granulated, with gas-vacuoles not attenuated; end cell broadly rounded.

Perceived floating in water bodies of paddy fields.

176. Lyngbya kuetzingii Schmidle

Plate-XII, Fig.176

Filaments straight or slightly curved or bent, rigid, single or many and gregarious, 2-3.5 µ broad, often short and only 30-70 µ long, with one end fixed or attached; sheath thin colourless, cells 1.5-2 µ broad, 1/2-1/3 as long as broad, not constricted at the cross-walls and not granulated, pale blue-green.

In soil cultures.

177. Lyngbya kuetzingiana Kirchner

Plate-XII, Fig.177

Thallus caespitose and lamellose, outside bright blue-green to olive-green, inside more or less colourless; filaments 3.6-5 µ broad, coiled; trichomes 3.5-4 µ broad, constricted at the cross-

walls only at the ends of the trichome, cross-walls granulated, cells as long as broad, blue-green; end cell obtuseor rotund.
In soil cultures.

178. Lyngbya lagerheimii (Mob.) Gomont

Plate-XII, Fig.178

Filaments single, or entangled with one another, irregularly spirally coiled or occasionally straight; sheath thin, colourless; trichome about 2 μ broad, cells 1.2 - 3 μ long not constricted at the cross-walls, with or without a single granule on either side, pale blue green; end cell rounded, not attenuated.
Perceived floating in water bodies of paddy fields.

179. Lyngbya lachneri (Zimmermann) Geitler

Plate-XII, Fig.179

Filaments attached by one end, single nearly straight, apparently without a sheath, but with a hyaline sheath visible under high magnification, seldom longer than 100 μ trichome 2.5 - 3.5 μ broad, constricted at the cross-walls, cells 1.5 - 3 μ long; end cell rounded, colour violet; hormogonia present.
In soil cultures.

180. Lyngbya limnetica Lemmermann

Plate-XII, Fig.180

Filaments straight or slightly curved or coiled, single, free-floating, 1 - 2 μ broad; sheath thin or narrow, colourless, cells 1 - 1.5 μ broad, quadrate to 1/3 rarely 1/8 as long as broad, 1 - 3 μ long, not constricted at the cross-walls, with or without a granule at the cross-walls, pale blue-green; end cells not attenuated, rounded.
In soil cultures.

181. Lyngbya lutea (Ag.) Gom.

Plate - XIII, Fig. 181

Thallus somewhat gelatinous, leathery, yellowish brown to olive-green, when dry often dark violet; filaments coiled and densely entangled; sheath colourless, smooth at first thin, but later up to 3 μ thick and lamellated, trichome not constricted at the cross-walls, not attenuated at the ends, 2.5 - 6 μ broad, olive-green, cross-walls granulated; cells quadrate to 1/3 times as long as broad, 1.5 - 5 μ long; and cells with rounded calyptra.

In soil cultures.

182. Lyngbya major Menegh ex Gomont

Plate-XIII, Fig. 182

Filaments long, straight forming dark-green caespitose bundles; sheath thick; colourless, lamellated, cells 11 - 16 μ broad, 1/4 - l/S as long as broad, 2-3.4 μ long, dull blue-green, granulated at the septa, slightly or not constricted at the cross-walls, end cell rounded with a slightly thickened membrane.

In soil cultures.

183. Lyngbya mamtensiana Menegh. ex Gomont

Plate-XIII, Fig.183

Thallus caespitose, blue-green, when dried violet, filaments long more or less flexible; sheath colourless, thick, outside rough; trichome 6-10 μ broad (rarely 13 u), not constricted at the cross-walls, cross-wall sometimes granulated, apices not attenuated, pale blue-green; cells 1/2-1/4 times as long as broad, 1.75-3.3 μ in length; end cell rotund, without calyptra.

In soil cultures.

184. Lyngbya martensiana Menegh, ex Gomont Var. **minor** Gardner

Plate-XIII, Fig.184

Filaments 250 - 500 µ long, slightly bent, 7.2 - 8 µ broad, sheath colourless, not lamellated, smooth, about 2 µ thick; trichome 4.8 - 5 µ broad, not constricted at the cross-walls, bright blue-green, granulated at the cross-walls; cells very short, 1.4 - 1.8 µ long; end cell capitate.

In soil cultures.

185. Lyngbya maharastrensis Kamat

Plate-XIII, Fig.185

Plant mass blue-green; filaments 12 - 13 µ broad, usually straight, but rarely showing regular to irregular spirals; sheath thick, unstratified, hyaline to yellow-brown; trichomes not tapering towards £he apices, not constricted at the cross-walls; cells much shorter than broad, 1/4 - 1/3 times as long as broad, 9 - 10 µ in diameter, 2 - 3.2 µ long; end cell rounded without calyptra.

In soil cultures.

186. Lyngbya majuscula Harvey ex Gomont

Plate-XIII, Fig.186

Thallus expanded, up to 3 cm long, dull blue-green to brown or yellowish brown; filaments very long, curved or seldom only slightly coiled; sheath colourless, lamellated up to 11 µ thick, outside often rough, trichome blue-green, brownish green or grey violet; not constricted at the crosswalls, not attenuated at the ends, 16-60 µ (or up to 80 u) broad, mostly 20-:40 µ broad; cells very short 1/6-1/5 times as long as broad, 2 - 4 µ long, cross-walls not granulated; end cells rotund, calyptra absent.

In soil cultures.

187. Lyngbya nordgardhii Wille

Plate-XIII, Fig.187

Thallus small, caespitose, grey violet or grey brown; filaments very long, straight, or more or less bent or curved (3-4 µ broad) attached to other filamentous algae; sheath thin, delicate, smooth, trichome grey violet, constricted at the cross-walls, 1.5 - 2 µ broad, not attenuated, cells 1 - 1/2 as long as broad, (up to 4 µ long); end cell rounded, without a thickened outer wall. Perceived aligned on paddy leaves beneath water.

188. Lyngbya palmarum (Martens) Bruhl et Biswas

Plate-XIII, Fig.188

Thallus extensive, densely tomentose, dark coloured, when dried pale blue-green, or pale yellowish; filaments creeping or ascending, suberect and flexuous, loosely or closely intertwined, with sheath 6 - 10 µ broad; sheath at first dull coloured later brownish, 0.6-1 µ thick; trichome not constricted at the joints, apices rotund, scarcely attenuated; cells 4.5 - 8 µ long (e: b: 8.8, 6.6, 8.6); contents densely granulated, septa rarely conspicuous, hidden by granules, septa not marked by series of granules; apical cell without a thickened outer membrane.
In cultures.

189. Lyngbya polysiphoniae Fremy

Plate-XIII, Fig.189

Filaments straight or curved, single or in bundle, epiphytic, up to 200 µ long; sheath very thin, delicate, papyraceous, colourless, trichome pale blue-green or violet, constricted at the cross-walls, about 2 µ broad, apices not attenuated; cells 1/2 - 1/3 as long as broad, cross-walls visible, not granulated; end cells convex, not capitate.
In cultures.

190. Lyngbya perelegans Lemm.

Plate-XIII, Fig.190

Thallus with many entangled, straight or curved filaments; filaments 1.5-2 µ broad; sheath thin, hyaline; trichomes 1-1.5 µ broad, not constricted at the cross-walls; cross-walls with a single granule on either side; cells 2-8 µ long, pale blue-green, end cell rotund, not attenuated.

Perceived aligned on paddy leaves beneath water.

191. Lyngbya putealis Mont. ex Gomont

Plate-XIII, Fig.191

Thallus caespitose, expanded, up to 1 cm long, penicillate, dull blue-green, when dry sometimes violet; filaments curved or nearly straight, at the base intricate, above more or less parallel; sheath thin, papyraceous, colourless, trichome 7.5 - 13 µ broad, distinctly constricted at the cross-walls, cross-walls sometimes granulated, not attenuated, not capitate, blue-green; cells quadrate of up to 1/3 as long as broad, 3-10 µ long; end cell rounded.

Perceived in water bodies of paddy fields.

192. Lyngbya rivularianum Gom.

Plate-XIII, Fig.192

Filaments single in the mucilage of other algae, coiled; sheath thin, colourless, papery; trichomes 0.75-0.8 µ broad, not constricted at the cross-walls, cross-walls not granulated, ends not attenuated pale blue-green; cells up to 4 times as long as broad, 2.3 - 3.2 µ long; end cell rotund; calyptra none.

Perceived in algal mats collected from wet soil surfaces.

193. Lyngbya stagnina Kuetzing.

Plate - XIII, Fig.193.

Thallus dull-green; filanents flexuous, 11-16 μ broad; sheath colourless or yellowish, lamellae 3, trichome 9.5-12 μ broad, not constricted at the cross-walls, distinctly granulated at the cross-walls; cells 2-4 μ long; end-cell broadly rounded, not attenuated.

In soil cul tures.

194. Lyngbya scotti Fritsch.

Plate - XIII, Fig.194.

Filaments mostly prominently coiled, occasionally nearly straight forming a thallus, 3.4-5 μ broad; sheath at first thin and colourless, later thicker, outer surface rough, unlamellated; trichome attenuated at the ends, not prominently constricted at the cross-walls, 2.6-3 μ broad; end cell prominently conical, not capitate, without calyptra; cells as long as broad or somewhat longer.

In soil cultures.

195. Lyngbya Subtilis West, W. var. **granulosa** Bruhl & Biswas.

Plate-XIV, Fig.195

Filaments solitary, flexuous, trichome 1.6 - 1.8μbroad, not constricted, pale blue-green, sheath close, colourless; cells 1.6 - 3.2μ long, contents densely granular.

In soil cultures.

196. Lyngbya schmidlei V. **major** Kamat.

Plate-XIV, Fig.196

Filaments straight or curved, epiphytic (3.5-) 3.8-4μ broad; sheath thin, hyaline; trichomes not constricted, at the cross-walls, not attenuated at the ends; cells shorter than broad, 3.5-3.Sμin diam. 2-2.5μlong; end cell rounded without calyptra.

In soil cultures.

197. Lyngbya seimplena (C. Ag.) J. Ag. ex Gomont

Plate-XIV, Fig.197

Thallus caespitose, seldom above 3 cm long, mostly dark yellowish green, sometimes when dry dark violet; filaments at the base decumbent, growing erect by their ends and entangled, curved or flexuous; sheath colourless, slightly slimy, mucilaginous, when old lamellated, up to 3 thick, trichome yellowish green or blue-green, not constricted at the cross-walls, cross-walls often granulated, at the ends slightly attenuated, capitate, 5 - 12μ commonly 7 - 10μ broad; cells 1/3 - 1/6 as long as broad, 2 - 3μ long; end cell with a flat conical or roundish calyptra.

In soil cultures.

198. Lyngbya shackletoni W. et G.S. West

Plate-XIV, Fig.198

Filaments nearly straight, 12 - 12.5μ broad; sheath firm, colourless, distinctly lamellatedt trichome not attenuated at the ends, 8.5 - 9.5μ broad, not constricted at the cross-walls; cells 2-2.4μ long, pale blue-green; end cell conical nearly as long as broad.

In soil cultures.

199. Lyngbya truncicola Ghose

Plate-XIV, Fig.199

Thallus thin, expanded, blue-green; filaments straight, more or less parallel, 14-16μ broad; sheath at first hyaline, and delicate, later firm and yellowish, unlamellated; trichome blue-green, 12 - 14μ broad, not constricted at the cross-walls, cell-walls not granulated; cells short, 3 4μ long, contents granular; apical cell rotund, not attenuated, calyptra none.

In soil cultures.

Genus - **Schizothrix** Kutz.

Trichome generally many, seldom single, in large soft to firm thallus, enclosed in a thin or thick sheath, often each trichome with its own sheath; sheath colourless to yellow-brown, red or seldom violet or blue in colour, its ends often pointed; filaments densely packed, forming a leathery stratus, occasionally only a trichome and sparsely branched, forming tuft-like growth or in erect bundles, which are more or less branched; thallus attached, seldom free-swimming.

200. Schizothrix arenaria (Berk.) Gomont

Plate-XIV, Fig.200

Thallus thin, not incrusted with calcium; filaments densely grouped, with branching at the tips, branches much contorted; sheath firm, at the ends pointed, outside even, in the lower portions thick and lamellated, with a few trichomes; trichome constricted at the cross-walls or unconstricted, 1.5 - 3μ broad, cells up to 5μ long; end cell pointed and conical.

In cultures as well as in collections from wet soils.

201. Schizothrix lardacea (Ces.) Gomont

Plate-XIV, Fig.201

Thallus expanding, not calcium incrusted, up to 3 cm thick, firm, elastic, lamellated, very sparsely branched; sheath colourless, firm, pointed at the ends, at first thin, later thick, even on the outside, with one or a few trichomes; trichomes not constricted at the cross-walls, cross-walls often granulated, 2 granules, 1.5 - 2μ broad; cells quadratic or slightly less longer than broad, 2 - 3μ long, blue-green; end cell obtuse.

In soil cultures.

202. Schizothrix lacustris A. Br. ex Gomont

Plate-XIV, Fig.202

Thallus cushion-shaped, or crustaceous, often free floating, without or with only a little calcium incrustation; filaments aggregated, at the ends richly branched; sheath thick, colourless or yellow, trichome many in a common sheath, in the older parts of the filaments and often spirally coiled, in the younger parts single, or a few together, constricted at the cross-walls (in herbarium specimens), 1-1.5µbroad; cells longer than broad, up to 4µlong.

Perceived in small water bodies of paddy field.

203. Schizothrix tenuis Woronich

Plate-XIV, Fig. 203

Filaments up to 3 mm long, and forming 30-50µ thick bundles, contorted, 2-4µ broad, trichomes 0.8-1.5µ broad, not constricted at the cross-walls; cells cylindrical.

In algal mat collections as well as in soil culture.

Genus - **Symploca** Kutzing

Trichome single in a thin sheath; filaments at first prostrate, later mostly forming erect bundles, partly false branched; sheath firm or later gelatinising; trichome straight, sometimes slightly attenuated; end cell not capitate, sometimes with a thickened outer membrane.

204. Symploca cartilaginea (Mont.) Gomont

Plate-XIV, Fig.204

Filaments arranged densely and parallel, forming bundles up to 1 cm high, dull blue-green; sheath seemingly thick, firm, cells 2-3µ broad; mostly longer than broad, 3 - 5.6µ long, not constricted at the cross-walls, blue-green; end cell obtuse with a very slightly thickened outer membrane.

Perceived on wet soil surfaces.

205. Symploca elegans Kutzing ex Gamont

Plate-XIV, Fig.205

Bundles erect, blue-green in colour, anastomosing; filaments bent, flexuous, parallel, agglutinated; sheath seemingly thick, trichome blue-green, 1.3 - 2μ broad, slightly constricted at the cross-walls, cross-walls not granulated; cells as long as broad or longer, 2 - 4μ long; end cell sub-conical, rounded, calyptra absent.

In soil cultures.

206. Symploca muralis Kutzing ex Gomont

Plate-XIV, Fig.206

Filaments very much bent, thickly arranged, erect up to 2 mm long, in blackish blue tufts; sheath firm, seemingly thick, cells 3.4 - 4μ broad, nearly quadratic, or up to 1/2 as long as broad, not constricted at the crosswalls; end cell obtuse conical, with slightly thickened outer membrane.

Perceived on much wet soil surfaces.

207. Symploca muscorum (Ag.) Gomont

Plate-XIV, Fig. 207

Filaments bent, forming a thin prostrate or creeping thallus, seldom erect, in bundles or forming a gelatinous Phormidum-like thallus, thallus blue-green or blackish; sheath thin, firm or more or less gelatinous, up to 2μ thick, trichomes not constricted at the cross-walls, cross-walls not granulated; cells 5-8μ broad, nearly quadratic or up to 2 times longer than broad, 5-11μ long; end cell mostly broadly rounded, with slightly thickened outer membrane.

In soil cultures.

208. Symploca parietina (A. Br.) Gomont.

Plate-XV, Fig.208

Filaments tortuous, forming small erect or prostrate anastomosing yellowish grey tufts; sheath seemingly thick, outer uneven, cells 1.8-3μ broad, longer than broad, 3.5 - 7μ long, yellowish green; trichome not constricted at the cross-walls, septa inconspicuous, rarely with 2 granules; end cell rounded, calyptra absent.

Perceived aligned to wet paddy stalks.

Genus - **Microcoleus** Desmazieres

Filaments unbranched or sparsely branched; sheath mostly colourless, more or less regularly cylindrical, not lamellated, sometimes when old, gelatinizing, trichomes very many in each sheath, densely aggregated, often coiled or contorted like a rope; ends straight, mostly attenuated; end cell more or less conical seldom capitate.

209. Microcoleus acutissimus Gardner

Plate-XV, Fig.209

Filaments nearly straight, 400-450 (-500)μ long, 20-35μ broad, with 15-30 trichomes; sheath colourless, uneven, more or less gelatinous; trichsmes mostly parallel, not constricted at the cross-walls, cross-walls frequently granulated, ends long, attenuated and pointed, 1.7-2.3μ broad; cells 2-4 times as long as broad, 3.5-8μ long, pale blue-green; end cell very acutely conical.

In soil cultures.

210. Microcoleus amplus Gardner.

Plate - XV, Fig.210.

Filaments 50-70μ broad, long, curved, containing 2-12 trichomes; sheath hyaline, not stratified; trichames not constricted at the cross-walls, attenuated at the ends; cells as long as broad, 4.5-5μbroad, 4-5μ long; end cell long, conical, without a cap or calyptra.
In soil cultures.

211. Microcoleus chthonoplastes Thurst ex. Gomont.

Plate - XV, Fig.2ll.
Filaments single, or forming an expanded dirty to dark green lamellated thallus, coiled, seldom branched; sheath sometimes gelatinizing, uneven, thick, having many closely grouped trichomes; trichome constricted at the cross-walls, 2.5 - 6μ broad, not granulated at the cross-walls; cells 1-2 times as long as broad, blue-green, 3.6 - 10μ long; end cells not capitate; pointed conical.
In soil cultures.

212. Microcoleus paludosus (Kutz.) Gamont:

Plate – XV, Fig.212.
Filaments single or forming a dark blue-green stratum, unbranched or sometimes divided at the end; sheath slightly gelatinous, with many straight or rope like trichomes; trichomes not granulated at the cross-walls, constricted at the cross-walls, 5-7μ broad; cells nearly as long as or twice as long as broad, 4-13μ long, bright blue-green; end cell not capitate, conical.
Perceived intermingled with algal filaments on wet soils.

213. Hydrocoleus subincrustaceus. Hansgirg.

Plate - XV, Fig.2l3.
Thallus dirty blue-green; filaments 15-25μ broad, with 1-2 trichomes; sheath hyaline; trichomes not constricted at the cross walls, not attenuated at the ends; cells longer than broad,

rarely shorter, 4.5-5.2μ broad, 4-7μ long; end cell rounded, without a cap or calyptra.
In soil cultures.

Genus - **Anabaenopsis** (Wolosz.) Mliller sensu strict.
Trichomes free-swimming, short, spirally coiled, like an Anabaena; heterocysts terminal and intercalary, intercalary ones in pairs formed by the unequal division of two adjoining vegetative cells; spores intercalary and formed away from the heterocyst.

214. Anabaenopsis arnoldii Aptekarj Var. **indica** Ramanathan
Plate-XV, Fig.214
Trichomes elongated, spirally coiled, coils ½ - 16 in number, spirals 30-80μbroad, trichome 7.6 - 11.4μ broad; cells short and barrel-shaped and slightly rounded between the cells, cells usually broader than long, but distinctly longer just before division, cells 7.6-(9.5)-11.4μ broad and 5.7 - (7.6) - 9.5μ long, contents blue-green in colour, and slightly granular with a few pseudo-vacuoles; heterocysts intercalary or terminal, origin of the heterocysts always intercalary in pairs, by the development of two cells derived by an unequal division of two adjoining vegetative cells; heterocysts spherical, 9.5-(13.3)-17.1μ in diameter, contents pale blue-green in colour, wall fairly thick with a distinct basal granule at one or even at both ends; spores ellipsoidal in shape and are 13.3-(15.2)-19.0μ broad, and 17.1-(19.0)-22.8μ long, spore-wall very thick and faintly striated, contents highly granular and greyish green in colour, multiplication by fragmentation of the trichome, usually between a pair of intercalary heterocysts.
In soil cultures.

215. Anabaenopsis circularis (G.S.West) Wolosz. et Miller

Plate-XV, Fig.215

Trichomes free-swimming, very short, mostly spirally coiled, with 1-1½ spirals, very seldom straight, 4.5 - 5µ broad; cells spherical or somewhat longer than broad, with a large granule; heterocysts spherical, 5 - 8µ broad, spores not known.

In soil cultures.

216. Anabaenopsis tanganyikae (West, G.S.) Wolosz. et Miller

Plate-XV, Fig.217

Trichome free-swimming, very short, spirally coiled, 1 - 2 (-3)µ mostly 1- ½ spirals, without sheath; trichomes not constricted at the cross-wall 2.4 - 2.5µbroad; cells cylindrical, 2 - 3 times longer than broad, 3.8 - 8.5µ long, without gas-vacuoles; heterocysts ellipsoidal, 3 x 5.5µ; spores ellipsoidal, single away from the heterocyst, 7µ broad; 13µ long, with colourless smooth membrane.

Perceived in water bodies of paddy fields.

Genus - **Cylindrospermum** Kutz.

Thallus mucilaginous, mostly dull blue green; trichome uniformly broad, short, without sheath, but in a common mostly very delicate and often imperceptible, mucilage of thin consistency; cells cylindrical, constricted at the cross-walls; heterocysts terminal, at both ends or at one end only, sometimes intercalary; spores single, rarely in series, next to the heterocyst on one side, much bigger than the vegetative cells.

217. Cylindrospermum gorakhpurense Singh, R.N.

Plate-XV, Fig.217

Trichomes blue-green, single, with deep constrictions at the joints, 3.5 - 5µ broad; cells cylindrical 7-16.5µ long;

heterocysts almost ellipsoidal one at each end of the trichome, spores ellipsoidal with rounded apex, 6-7.3μ broad, and 13.2-16.5μ long, spores sub-terminal at either end of the trichome, with a thick yellowish brown outer membrane provided with a delicate needle-shaped projections, 13.2-20μ broad and 26.4-33μ long, without exospore 13.2-14μ broad.
Perceived in algal mats on wet soils surfaces.

218. Cylindrospermum indicum Rao, C.B., orth. Mut. De. Toni

Plate-XV, Fig.218
Trichome single with deep constrictions at the joints, 3.7μ broad, dark blue-green; cells almost quadrate, or more or less barrel-shaped, 3-4.5μ long; heterocysts spherical, subspherical, subconical, or ellipsoidal, sometimes cylindrical, one at each end of the trichome, 2.3-5.8μ broad and 3-7.6μ long; spores almost ellipsoidal, sometimes cylindrical, subterminal at either end of the trichome, with a thick yellowish brown outer membrane possessing a smooth outer margin and exhibiting faint lines running across in optical section, without membrane 8.8-9μ broad and with membrane 10-12μ broad, without membrane 15-18.5μ long and with membrane 18-22μ long.
Perceived on the edges of water bodies in paddy fields.

219. Cylindrospermum licheniforme Kutz ex. Born. et. Flah.

Plate-XV, Fig.219
Thallus mucilaginous, orbicular, confluent, later more or less expanded, blackish green; trichome (2.5-) 4.2μ broad, constricted at the cross-walls, pale blue-green; cells quadrate to cylindrical, 4-5μ long; heterocysts oblong, (4-) 5-6μ broad, 7-12μ long; spores oblong to ventricose elliptic, 12-14μ broad, 20-30 (-38)μlong, with a smooth reddish brown epispore.

In soil cultures as well as in algal mat collections from much wet soil surfaces.

220. Cylindrospermum majus Kutzing ex. Borm et. Flah.

Plate-XV, Fig.220

Thallus expanded, mucilaginous, blackish green; trichome 4-5μ broad, constricted at the cross-walls, light blue-green; cells cylindrical, 5-6μ long; heterocysts oblong, somewhat broader than the trichome, up to about 10μlong; spores ellipsoidal, 10-15μ broad, 20-30 (-38)μ long, epispore brownish with distinct papillae.

Perceived in much wet soil surfaces as well as in small water bodies.

221. Cylindrospermum musicola Kutzing ex. Borm et. Flah.

Plate-XV, Fig.221

Thallus expanded, mucilaginous, blackish-green; trichomes 3-4.7μ broad, constricted at the cross-walls, light blue-green; cells 4 (-5)μ long, cylindrical, or nearly quadrate; heterocysts oblong, 4(-5)μ broad, 5-7μ long; spores oval, 9-12μ broad, 10-20μ long, epispore smooth, yellowish brown.

In soil culture as well as in algal mat collections from paddy fields.

222. Cylindrospermum musicola Kutzing ex. Borm et. Flah. Var. **longispora** Dixit

Plate-XV, Fig.222

Trichome 3.7-4μ broad, 4.5 - 7.5μlong; heterocysts 5-6μ broad and 7.5-10μ long; spores 7.5-12.5μ broad and 21.5 - 31.5μ long.

Perceived in water bodies in paddy fields as well as in soil cultures.

223. Cylindrospermum musicola Kutzing ex Born. et Flah. Var. **kashmiriensis** Bharadwaja

Plate-XVI, Fig. 223

Thallus thin, velvety, shining, bluish-green, irregular layer, very soft and mucilaginous, attached, up to 1.5 cm wide; trichomes in the upper part of the thallus more or less straight and parallel, in the lower parts irregularly bent, and more or less entangled, 125 to 200μ long, slightly constricted at the joints; cells barrel-shaped, twice as long as broad, 2.6-3.9μ broad, and 2.6-8.4μ long, septa indistinct; heterocysts one at either end in mature filaments or only at one end in younger ones, oval or ellipsoidal, 3.9-5.2μ broad and 5.2-10.5μ long, intercalary heterocysts single or in pairs, shorter or longer than vegetative cells; spores adjoining to the terminal heterocysts, barrel-shaped or ellipsoidal, 5.2-7.8μ broad and 9.4-13.6μ long, epispore thick, smooth and brownish.

Perceived in water bodies in paddy fields.

224. Cylindrospermu n michailovskoense Elenkin

Plate- XVI , Fig. 224

Thallus expanded, blue-green; trichome pale blue-green, mostly bent, sometimes straight, aggregated; cells quadrate or mostly cylindrical, more or less constricted at the cross-walls, 3.5 - 5μ broad, 6-7μ long; heterocyst nearly spherical, oblong, 5-6μ broad, 7-8μlong; spores single, broadly ellipsoidal, or oblong, 8-13μ broad, 12-32.6μ long, epispore colourless.

Perceived on the edges of water bodies in paddy fields.

Genus - **Nostoc** Vaucher

Thallus mucilaginous, gelatinous or coriaceous, first globose to oblong, later globose, foliose, filiform, bullose, solid or hollow, free or attached, the periphery dense and darkly coloured;

filaments flexuous, curved or entangled; sheath sometimes distinct, generally diffluent; trichome torulose; cells depressed, spherical, barrel-shaped or cylindrical; heterocysts intercalary, and in young condition terminal; spores spherical, or oblong, formed centrifugally in series in between the heterocysts.

225. Nostoc coeruleum Lyngbye ex Born. et Flah.

Plate-XVI, Fig.225

Thallus small, up to 1 cm diam., globose, solid, pale blue-green, seldom brownish, solitary or aggregated, not infrequently proliferated, with a firm outer layer; filaments densely entangled, flexuous; sheath mostly indistinct; trichome 5-7μ broad, torulose; cells short, barrel-shaped or subspherical; heterocysts subspherical up to 10μ broad; spores not known. Perceived aligned to submerged portions of paddy as well as in soil cultures.

226. Nostoc commune Vaucher ex Born. et Flah.

Plate-XVI, Fig.226

Thallus firm, gelatinous, at first globose, later flattened, expanding, undulated, membranous or leathery, sometimes irregularly torn, often perforated, many centimeters diam., blue-green, olivaceous or brown; filaments flexuous, entangled; sheath mostly distinct only at the periphery, thick, yellowish brown, often lamellated, inside the thallus more or less distinct, but hyaline; trichome 4.5-6μ broad, cells short barrel-shaped or nearly spherical, mostly shorter or a little longer than broad, 5μ long; heterocysts nearly spherical, about 7μ broad; spore only once observed, as big as the vegetative cells epispore smooth colourless.

Perceived on moist soil surfaces along with other algae as well as in soil cultures.

227. Nostoc calcicola Brebisson ex Born. et Flah.

Plate-XVI, Fig. 227

Thallus mucilaginous, slightly diffluent, expanded, olive, grey, or blue green, often up to 5 cm in diam., filamently loosely entangled; sheath mostly indistinct, or distinct only at the periphery of the thallus, colourless or yellowish brown; trichome 2.5μ broad, pale blue-green; cells barrelshaped, subspherical, rarely longer than broad; heterocysts subspherical, 4 (-5) μ broad; spores subspherical, 4-5μ broad, with smooth yellowish membrane.

Perceived aligned to wet parts of paddy as well as in soil cultures.

228. Nostoc ellipsosporm (Desm.) Rabenh. ex Born.et Flah.

Plate-XVI, Fig. 228

Thallus gelatinous, irregularly expanded, attached by the lower surface, reddish brown; filaments flexuous, loosely entangled; trichome about 4μ broad, light bluegreen or olivaceous; cells cylindrical, 6-14μ long; heterocysts subspherical, or oblong, 6-7μ broad, 6-14μ long; spores ellipsoidal to oblong cylindrical, 6-8 μ broad, 14-19μ long, epispore smooth, hyaline or brownish.

Perceived floating in small water bodies in paddy fields.

229. Nostoc entophytum Born. et Flah.

Plate-XVI, Fig.229

Thallus small macroscopic inconspicuous, blue-green or yellowish, growing on aquatic plants, also in the cells or lacuna of plants; filaments densely entangled; sheaths generally distinct, hyaline later brownish; trichomes 2.5 - 3μ broad, torulose; cells short, barrel-shaped; heterocysts broader than the vegetative

cells; spores spherical or slightly compressed, 5 - 5µ broad, rarely oblong, 5-8µ long, with brown smooth epispore.
Perceived on wet soil, as well as in soil cultures.

230. Nostoc humifusum Carmichael ex Born. et Flah.

Plate-XVI, Fig. 230
Thallus gelatinous or mucilaginous, irregular with variable size, sometimes punctiform, sometimes many smaller ones uniting and expanding, tuberculate; olivaceous or brownish in colour, attached by the under surface; filaments twisted and flexuous, densely entangled; sheath mostly indistinct, rarely distinct and yellowish brown; trichomes 2.2-3µ broad, more or less blue-green; cells subspherical or nearly twice as long as broad; heterocysts 3µ broad, subspherical; spores subspherical or oval, 4µ broad, 6µ long, epispore smooth and yellowish.
Perceived in water bodies along with other algal members in paddy fields.

231. Nostoc linckia (Roth) Bornet ex Born. et Flah.

Plate-XVI, Fig. 231
Thallus varying in size, sometimes punctiform, sometimes tuberculate, at first globose later irregularly expanding, torn, gelatinous, blue-green to violet, or blackish green or brown; filaments densely entangled, flexuous or highly coiled; sheath diffluent and colourless inside, distinct only in the peripheral portion; trichomes 3.5-4µ broad, pale glue-green; cells short barrel-shaped; heterocysts subspherical; spores subspherical, 6-7µ broad, 7-8µ long, epispore smooth.
Perceived in algal collections from paddy fields as well as in soil cultures.

232. Nostoc microscopicum Carm. ex Born. et Flah.

Plate-XVI, Fig. 232

Thallus spherical or ellipsoidal, about 1 cm diam., or only very seldom larger, soft, but with a firm outer surface, first glistening later olivaceous or brown; filaments loosely entangled; sheath more or less distinct, yellowish; trichome 5-8µ broad, blue-green, or olive-green; cells subspherical or barrel-shaped; heterocysts nearly spherical, 7µ broad; spores oval, 6-7µ broad, 9-15µ long, olivaceous, epispore smooth.
Perceived aligned to wet paddy surfaces as well as in soil cultures.

233. Nostoc muscorum Ag. ex Born. et Flah.

Plate-XVI, Fig. 233
Thallus gelatinous-membranous, irregularly expanded, attached by the lower surface, tuberculate, dull olive or brown, 2-5 cm diam.; filaments densely entangled; sheath distinct only at the periphery of the thallus, yellowish brown; trichome 3-4 (-5)µ broad; cells short barrel-shaped to cylindrical, up to twice as long as broad; heterocysts nearly spherical, 6-7µ broad, spores oblong, many in series, 4-8µ broad, (u-) 8-12µ long, epispore smooth and yellowish.
Perceived on moist soil surfaces along with other algal members.

234. Nostoc maculiforme Born. et Flah.

Plate-XVI, Fig. 234
Thallus flat attached, forming thin irregular blue-green spots; filaments flexuous, contorted densely packed, into loose net-like masses; sheath distinct, close; trichome 3.5-4µ broad, cells oblong to depressed spherical; spore (immature) globose, mostly 6µ rarely 8µ broad, in long catenate chains, exposure smooth, thick and fused with the sheath.
In soil cultures.

235. Nostoc paludosum Kutzing ex Born. et Flah.

Plate-XVI, Fig. 235

Thallus microscopically not visible, punctiform, gelatinous; sheath broad, colourless or yellowish brown; trichomes 3-3.5µ broad; cells as long as broad, barrel-shaped, pale blue-green; heterocysts broader than the vegetative cells; spores oval, 4-4.5µ broad, 6-8µ long, with smooth colourless membrane.

Perceived in water bodies of paddy fields as well as in soil cultures.

236. Nostoc piscinale Kutzing ex Born. et Flah.

Plate-XVII, Fig. 236

Thallus at first globose, later bullose, or variously tuberculate, mucilaginuous or gelatinous, light blue-green, olive-green to brown; filaments flexuous, loosely entangled; sheath distinct at the periphery of the thallus, brownish, diffluent in the inner portions of hyaline; trichome 3-7µ broad; cells shorter or longer than broad up to nearly twice as long as broad; heterocysts subspherical or oblong, 4.5-6µ broad; spores globose, 6-7µ broad, in long chains, epispore smooth and hyaline with a sheath.

Perceived on wet soil surfaces of paddy fields.

237. Nostoc punctiforme (Kutz.) Harjot

Plate-XVII, Fig. 237

Thallus sub-globose up to 2 mm diam., scattered or confluent, attached; filaments flexuous, densely entangled; sheath delicate, hyaline, mucous; trichome 3-4µ broad, cells short barrel-shaped or ellipsoidal, blue-green; heterocysts 4-6.5µ broad; spores subspherical, or oblong, 5-6µ broad and 5-8µ long, epispore thick and smooth.

Perceived in water bodies of paddy fields along with other algal members.

238. Nostoc spongiaeforme Agardh ex Born. et Flah.

Plate-XVII, Fig. 238

Thallus at first globose, gelatinous, later expanding, verrucose, bullose, light blue-green, violet or brownish, filaments flexuous, loosely entangled; sheath diffluent in the inside, more or less distinct along the periphery, yellowish brown; trichome about 4µ broad, blue-green to violet; cells partly cylindrical and up to 7µ long, partly short barrel-shaped; heterocysts subspherical or oblong, 7-8µ broad, somewhat longer than broad; spores oblong, 6-7µ broad, and 10-12µ long, epispore smooth, at first colourless, later yellowish.

Perceived floating in water bodies in paddy fields.

239. Nostoc sphaericum Vaucher ex Born. et Flah.

Plate-XVII, Fig. 239

Thallus free, globose, 1-15 mm diam., later irregularly plicate-tuberculate, thick, sometimes 6-7 cms in diam., olive-green, yellow or violet brown, with firm outer layer; filaments flexuous, densely entangled; trichome 4 rarely 5µbroad; cells compressed spherical or barrel-shaped; heterocysts (4-) 6µ broad, sub-spherical; spores oval 5 (4-6)µ broad, 7 (6-8)µ long, epispore thick and brownish.

In soil cultures.

Genus - **Anabaena** Bory

Trichomes uniformly broad throughout or apices alone somewhat attenuated, sheath absent or more or less diffluent, forming a free, torn or floccose or soft mucilaginous thallus; heterocysts generally intercalary; spores single or in long series, formed from near the hetero-cysts or in between the heterocysts.

240. Anabaena anomala Fritsch

Plate-XVII, Fig. 240

Thallus thin, gelatinous, blue-green; trichome commonly densely or irregularly aggregated, more or less contorted, moniliform, apical cell rounded, or subconical, apex obtuse; cells generally 2-3µ broad, as long as broad, spherical or barrel-shaped, sometimes up to 1½ times as long as broad, about 3.5µ (in immature spores?); heterocysts rare, commonly single, intercalary, spherical, 3.5-5.5µ broad, commonly isolated, exospore solid, endospore thin.

Perceived on wet soil surfaces as well as in soil cultures.

241. Anabaena ambiqua Rao, C.B.

Plate-XVII, Fig. 241

Trichomes free or completely enclosed in a mucilaginous envelope or sheath; in the latter case occurring singly or in group; trichomes without sheaths free-floating, ensheathed trichomes free-floating or attached to host by one end, generally occurring single, but occasionally in dense clusters; sheath usually firm, hyaline, with smooth or rough outline; usually 300-500µ (rarely up to 1 mm) long, 10-50µ broad; trichomes straight, or bent, usually slightly tapering at the ends, with end cells having rounded apices; usually 250-300µ long, sometimes longer; cells barrel-shaped with deep constrictions at the joints, septa indistinct, 4.9-6.6µ broad, 3.5 - 5µ long, cell contents deep blue-green, and coarsely granular; heterocysts at intervals, almost spherical, sometimes with slightly flattened ends, usually broader than the cells, 6.4-9µ (rarely up to 10 µ) in diam.; spores formed in free trichomes as well as in those enclosed in a sheath one on each side of the heterocyst, usually ellipsoidal sometimes cylindrical, with rounded ends or occasionally

flattened ends, 8.4-10.9μ broad and 13.3-16.2μ long, exospore thick and hyaline, endospore thin and transparent.
In soil cultures.

242. Anabaena circunalis Rabenhorst ex Born. at Flah. Var. **crassa** Ghose

Plate-XVII, Fig. 242
Trichome free-swimming, single, semi-circular, loosely coiled up to 4 times; cells nearly spherical, but generally shorter than broad, 5-7μ in diameter, with pseudovacuoles; heterocysts globose, up to 8μbroad; spores not seen.
In soil cultures.

243. Anabaena doliolum Bharadwaja

Plate-XVII, Fig. 243
Plant mass mucilaginous, pale blue-green; trichome single, free-swimming, straight, curved or slightly coiled, 3.6-4.2μ broad, slightly tapering at the ends, with conical apical cell, possessing almost pointed apex, cells barrel-shaped, as long as broad or a little longer or shorter than broad; heterocysts barrel-shaped, 5.2 - 6.3μ broad and 6.3 - 9.4μ long; spores ellipsoidal, with almost pointed apices in short or long chains, adjoining the heterocysts but developed centrifugally, epispore thick, smooth and hyaline or yellow brown, 4.2-6.2μ broad and 6.3-11.5μ long.
Perceived in water bodies aligned with other floating algal members.

244. Anabaena fertilissima Rao, C.B.

Plate-XVII, Fig. 244
Trichome single, straight or bent, with almost rounded end cells, up to 350μ long, 5-5.6μ broad, at the apex 4μ broad; cells barrel-shaped, 4.8-8μ long; heterocysts almost spherical,

6.4-8.4μ broad; spores in long chains, often making the whole trichome sporogenous, adjoining the heterocysts but formed centrifugally, almost spherical, with a smooth hyaline outer wall, 4.8-8μ broad and 3.2-8.8μ long.

In soil cultures.

245. Anabaena Iyengarii Bhəradwaja

Plate-XVII, Fig. 245

Trichome single or irregularly curved, 5.2-6.3μ broad, end-cell conical with rounded apex; cells barrel-shaped, as long as broad, or slightly shorter or longer than broad; heterocysts barrel-shaped, rarely spherical, 7.3-8.4μ broad and 7.3-10.5μ long; spores ellipsoidal often in long or short chains, rarely single on both sides of the heterocysts, 8.4-10.5μ broad and 10.5-21μ long, epispore thick, smooth and yellowish brown.

Perceived aligned with algal members in water bodies of paddy fields.

246. Anabaena Iyengarii Bharadwaja Var. **tenuis** Rao, C.B.

Plate-XVII, Fig. 246

Plant mass floccose, thin, free-floating, pale blue-green; trichomes single, straight or irregularly curved, 3.5 - 4.5μ broad, end cells conical with rounded apices; cells barrel-shaped, as long as broad, or slightly shorter than or longer than broad, (2.5-) 3-6.4μ long; heterocysts more or less barrel-shaped, sometimes sphetical, 4.8-6.4 (-8)μ broad and 5.2-9 (-12)μ long; spores ellipsoidal, or cylindrical, with rounded ends, single or in pairs on either side of a heterocyst, 7.5 - 9.6μ rarely 10.5μ broad and 9-19μ rarely 21 (-24)μ long, with a smooth hyaline outer wall.

Perceived on moist soil surfaces as well as in soil cultures.

247. Anabaena laxa (Rabehn.)

Plate-XVII, Fig. 247

Thallus floccose, free-floating or attached to other algae, blue-green; trichomes 4-5µ broad, straight, parallel, sometimes free and sometimes with a mucilaginous sheath; cells spherical or barrel-shaped, 5-6µ long, apices hardly attenuated, end-cell rounded; heterocyst spherical, 6µ broad, or elongate and up to 10µ long; spores single or many away from the heterocyst, 6-8µ broad, and 14-17 (-20)µ long; epispore smooth and yellowish.

Perceived in water bodies of paddy fields as well as in soil cultures.

248. Anabaena naviculoides Fritsch

Plate-XVII, Fig. 248

Thallus thin, gelatinous, blue-green; trichome elongate, more or less coiled, moniliform, apices acuminate; cells 3.5 - 5µ broad, as long as or shorter than broad (developing akinetes), apical cell obtuse conical, or acute, sometimes retuse; heterocysts rare, intercalary, single, barrel-shaped, 5-6µ broad, as long as or slightly longer than broad; spores in long series, irregularly disposed, ellipsoidal, ends acute and protracted, exospore thin, hyaline, more or less 11-12.5µ long, 6-7µ broad and 8.5-9µ long x 5.5µ broad.

In soil cultures as well as in algal collection from wet soils of paddy fields.

249. Anabaena oryzae Fritsch

Plate-XVIII, Fig. 249

Thallus soft, green, gelatinous, membranous, trichomes short, straight, densely aggregated, generally parallel cells 2.5-3µ broad, more or less barrel-shaped, 1½ - 2 times as long

as broad; heterocysts terminal and intercalary, broader than the vegetative cells, 3-3.5μ broad, terminal ones conical and twice longer than broad, intercalary ones, single, or 2-3 in series, generally barrel-shapad, sometimes spherical, single; spores rarely single next to the terminal heterocyst, commonly away from the intercalary heterocysts, single or 2-7 in series, subspherical or short ellipsoidal, 5 x 5, 6.5 x 7, 5 x 5.5, 5 x 6.5, 5 x 6 μ, exospore yellowish brown.
Perceived intermingled with other algae forming a algal mat on wet soil surfaces as well as in soil cultures.

250. Anabaena oscillariodes Bory ex Born. et Flah.

Plate-XVIII, Fig. 250
Thallus gelatinous, dark green, trichome 4.2-6 (generally 5-6)μ broad; cells barrel-shaped as long as broad, or somewhat longer or shorter than broad, end cells rounded; heterocysts spherical or oval, 6-8μ broad, 6-10μ long; spores on both sides of the heterocysts; single or 2-3, at first oval, later rounded - cylindrical, 8-10μ broad (more commonly 10 μ), 20-40μ long, epispore smooth and pale brown.
Perceived in small water bodies of paddy field as well as in soil cultures.

251. Anabaena oscillariodes Bory ex Born. et Flah. Var. **angustus** Bharadwaja

Plate-XVIII, Fig.251
Trichomes single, irregularly bent or spirally coiled 4.2-5.2μ broad; end cell rounded; cells barrel-shaped, as long as or slightly shorter or longer than broad; heterocysts intercalary, very rarely terminal, ellipsoidal 5.2-6.3μ broad and 7.3-10.5μ long; spores long, cylindrical, single or in short or long chains,

on both sides of the heterocysts, 6.5-8.4μ broad and 14.7-41.0μ long, epispore smooth yellow-brown.
Perceived free floating along with other algal members in water bodies of paddy fields.

252. Anabaena orientalis Dixit

Plate-XVIII, Fig. 252
Trichomes singlet straight or slightly curved, 2.5-4μ broad, cells quadrate or cylindrical, rarely slightly barrel-shaped, up to twice (or even thrice) as long as broad, 3.7-4.8μ long, end-cell conical with rounded apex; heterocysts singlet intercalary or cylindrical or slightly ellipsoidal with rounded end-walls, 4.8-5.5μ broad, and 7.4-9.2μ long; spores one on each side of a heterocyst, ellipsoidal, 7.4-9.2μ broad and 14.8-16.6μ long.
In soil cultures.

253. Anabaena randhawe Venkataraman, G.S.

Plate-XVIII, Fig. 253
Trichomes more or less straight, with occasional terminal spherical heterocyst at one end of the trichome; cells 4.0 - 4.5μ broad, 3.8 - 6.0μ long, heterocysts barrel shaped to spherical, 5.0 - 5.7 x 5.0 - 7.1 μ; spores away from the heterocyst, 5.7 - 8.5μ in diam.
In soil cultures.

254. Anabaena sphaerica Bornet et Flahault

Plate-XVIII, Fig. 254
Thallus floccose; blue-green; trichomes moniliform, straight, arranged parallel, 5-6μ broad, with an indistinct mucilaginous sheath; cells spherical to short barrel-shaped; end cells rounded; heterocysts subspherical, 6-7μ broad; spores on one or both sides of the heterocysts, subspherical to oval, 8-12μ broad,

12-18μ long, one to few together, epispore, smooth, yellowish brown.
Perceived in water bodies of paddy fields as well as in soil cultures.

255. Anabaena sphaerica Bornet et Flahault Var. **attenuata** Bharadwaja

Plate-XVIII, Fig. 255
Thallus floccose, gelatinous, thin, free-floating, pale blue-green; trichomes blue-green, curved or straight, more or less entangled with each other, 3.2-5.2μ broad, slightly attenuated at the ends, with rounded end-cells, without a mucilage sheath; cells spherical or slightly barrel-shaped, 3-6μ long; heterocysts intercalary, or slightly pressed from both sides, (4-) 5.2-6.2 (-8)μ diam.; spores single (or two together) on one or both sides of the heterocyst, spherical or oval, 10.5-12.6μ broad and 10.5-14.7 (-16.5)μ long, with smooth yellow outer wall.
Perceived along the banks of water bodies in paddy fields.

256. Anabaena torulosa (Carm.) Lagerh. ex Born. et Flah.

Plate-XVIII, Fig. 256
Thallus mucilaginous, thin, blue-green; trichome 4.2-5μ broad, apical cell acutely conical; cells barrel-shaped, as long as or somewhat shorter than broad; heterocysts subspherical or ovoid, 6μ broad and 6-10μ long; spores on both sides of the heterocysts, developed centripetally, single or many, subcylindrical with rounded ends, sometimes constricted in the middle, 7-12μ broad, up to twice as long as broad, epispore smooth and pale brown in colour.
Perceived aligned on the submerged portions of paddy as well as in soil cultures.

257. Anabaena vaginicola Fritsch at Rich

Plate-XVIII, Fig. 257

Trichomes many in a common mucilaginous sheath, rarely single, more or less parallel, 4-4.5μ broad, single trichome with sheath 11.5μ broad, many trichomes in a sheath, 17-21μ broad, attached to other plants; cells sub-quadrate, so elongate cylindrical; apical cell acuminate conical; heterocysts cylindrical, 4-5μ broad, 6-10μ long, spores short cylindrical or oblong, contiguous with the heterocysts, often 4-5 in series, wall hyaline, 6.5 - 10μ broad and 12-17.5μ long.

In soil cultures.

258. Anabaena variabilis Kutzing ex Born. et Flah.

Plate-XVIII, Fig. 258

Thallus gelatinous, dark-green; trichome without any sheath, flexuous, 4-6μbroad, more often 4.2-5μbroad slightly constricted at the cross-walls, end-cells conical, obtuse; cells barrel-shaped, sometimes with gas-vacuoles, 2.5-6μlong; heterocysts spherical or oval, 6μbroad, up to 8μlong; spores formed centrifugally, not contiguous with the heterocysts, barrel-shaped, in series, 7-9 (-11)μbroad, 8-14μlong, epispore smooth, or with fine needles, colourless or yellowish brown.

Perceived on wet soil surfaces as well as in soil cultures.

259. Anabaena variabilis Kutzing ex Born. et Flah. Var. **ellipsospora** Fritsch

Plate-XVIII, Fig. 259

Thallus thin, gelatinous, blue-green, trichome slightly flexuous, loosely aggregated, constricted at the cross-walls, apical cells rounded, sometimes conical or obtuse; cells 4-4.5μ broad, generally elongate, 5.5-9μ long, cylindrical or slightly barrel-shaped, sometimes later isodiametrical and torulose

or moniliform; heterocysts intercalary, solitary, spherical or barrel-shaped, 5.5-6µ broad, and 5-8µ long, heterocysts rarely terminal, hemispherical; spores in catenate series, away from the heterocysts, ellipsoidal, more or less oblong, 5-7µ broad and 7.5-12µ long, endospore thin, expspore separated from the endospore, indurated over a part of its surface only.
In soil cultures.

260. Anabaena variabilis Kutzing ex Born. et Flah. Var. **kashiensis** (Bharadwaja)

Fritsch
Plate-XVIII, Fig. 260
Thallus dense, soft, mucilaginous, deep green; trichomes 3.1-4.2µ broad, blue-green, often irregularly curved and more or less entangled with each other, slightly constricted at the joints, attenuated at the ends, the terminal cell being conical with a sharp or rounded apex, without mucilaginous sheath; cells cylindrical, up to twice as long as broad, rarely barrel-shaped, and almost as long as broad; heterocysts single intercalary, and distributed at regular intervals throughout the length of the trichome, cylindrical, 4.2-5.2µ broad and 8.4-12.6µ long; spores in short or long chains, ellipsoidal or barrel-shaped, remote from the heterocyst, 4.2-6.3µ broad and 6.3-10.5µ long, with a thick, smooth and colourless outer wall.
Perceived floating on water bodies of paddy fields as well as in soil cultures.

Genus - **Pseudanabaena** Lauterborn

Trichome single, not forming a thallus, with cells distinctly placed from one another, with brisk creeping movements, without sheath; cells cylindrical and at the ends rounded or oval.

261. Pseudanabaena catenata Lauterb

Plate-XIX, Fig. 261
Cells cylindrical, at both ends truncated, blue-green, sometimes somewhat brownish, 2μ broad and 3μ long.
Perceived free floating in water bodies of paddy fields.

262. Pseudanabaena schmidlei Jaag. 0

Plate-XIX, Fig. 262
Trichomes 6-7μ broad, up to 300μ long; cells more or less as long as broad, after division about ½ diam.
In soil cultures.

263. Pseudanabaena schmidlei Jaag. O f. **gracilis** Skuja

Plate-XIX, Fig. 263
Trichome 3.5-4μ broad, and 50-136μ long.
In soil cultures.
Sub family - Aulosirae Bornet et Flahault

Genus - **Nodularia** Mertens

Filaments free with a sheath; trichome in vegetative condition extremely uniform in breadth; sheath hyaline, delicate, mucilaginous, sometimes diffluent; cells short, depressed, discoid; heterocysts intercalary, compressed; spore globose, subglobose to discoid, formed singly or in series between the heterocysts, epispore smooth.

264. Nodularia spumigena Mertens ex Born. et Flah. Var. **major** (Kutz.) Born. Et Flah.

Plate-XIX, Fig. 264
Filaments 12-18μ broad; spores depressed, elliptical 14-15μ broad and 6-7μ long.

Perceived in algal collections from water bodies in paddy fields.

Genus - **Aulosira** Kirchner

Filaments free, sparse or in fascicles, generally uniformly broad, without differentiation of base and apex; trichomes with sheath, indefinite; heterocysts intercalary; spores often in series, formed near a heterocyst or away from it, cylindrical.

265. Aulosira aenigmatica Fremy

Plate-XIX, Fig. 265

Thallus expanded, woolly, dark blue-green floating; filaments more or less intricate, suberect, 6.5-8µ broad; sheath colourless, close, membranous, thin, not coloured violet with chlor-zinc-iodide; trichome pale blue-green, 5.5-6µ broad, more or less constricted at the cross-walls; cells subquadrate, longer or shorter than broad, contents finely granular, apical cells attenuated, obtuse conical; heterocysts as broad as or slightly broader than the trichomes oblong to spherical, single or rarely two; spores variously situated, in series, about 7µ broad, barrel-shaped or compressed-spherical, commonly shorter than broad, epispore colourless and weakly thick (immature?).

Perceived floating in water bodies of paddy fields.

266. Aulosira fertilissima Ghose

Plate-XIX, Fig. 266

Stratum expanded, dark blue-green, membranous; trichomes straight or a little flexuous, parallel or densely intricate, rarely with very short pseudobranches; cells (4-) 6-11µ broad and (5-) 7-10µ long, cylindrical when young, later barrel-shaped, contents granular; sheath thick, at first gelatinous and hyaline, later firm and brown; heterocysts intercalary, oblong or elliptical, 8-9µ broad and 10-14µ long; spores in series usually alternating

with dead cells, generally oblong-elliptical, sometimes angular due to compression 10-13μ broad and 18-24μ long.
Perceived aligned to fallen leaves of paddy in water bodies.

267. Aulosira fertilissima Ghose Var. **tenuis** Rao, C.B.

Plate, XIX, Fig. 267

Plant mass fibrous and greyish blue; filaments free, (4.8-) 5.6-6.4μ broad, more or less straight, and closely crowded together; sheath thin, firm and brown, 0.3-0.8μ thick, sometimesenveloped by diffluent mucilage; trichome slightly tapering at the ends 3.3-4.8 (-5.5) μ broad, rarely constricted at the cross-walls; end-cell with a rounded apex; cells cylindrical, sometimes quadrate 3.3-19.8 long; heterocysts cylindrical, with rounded endwalls broader than the trichome and causing the filaments to bulge, 4.5-7μ broad and 8.2 - 19.8 (-23.2)μ long; spores in long chains, ellipsoidal to cylindrical, 4.8-8.8 (-10)μ broad and 10.5 - 19.2μ long, intercalated by moribund cells.
In soil cultures.

268. Aulosira implexa Bornet et Flahault

Plate-XIX, Fig. 268

Filaments 5-10 mm long, straight or curved, often fasciculate; sheath thin, membranous, hyaline; trichome 8-9μ broad; cells quadrate, cylindrical somewhat longer or shorter than broad, blue-green, slightly constricted at the cross-walls granular; heterocysts quadrate to oblong; spores 4-32 in series, 8-9μ broad, 16-34μ long, epispore smooth.
In soil cultures.

269. Aulosira prolifica Bharadwaja

Plate-XIX, Fig. 269

Thallus forming a dense mucilaginous scum at first pale brownish green to blue-green or green; filaments 4.8-5.6μ broad, parallel, straight or slightly curved, becoming brownish or greenish yellow; sheath thin, first homogenous, later with an outer diffluent portion up to 1.75μ thick, and inner uniformly thick, 1μ somewhat firm; trichomes markedly constricted at the cross-walls, apical cells tapering to the rounded apex; cells cylindrical, generally longer than broad, 3.1 - 5.2μ broad and 6.3 - 21.0μ long, rarely quadrate; heterocysts intercalary or terminal, broader than the trichome, bulging out the sheath, intercalary heterocysts generally single, rarely in pairs, regularly placed, ellipsoidal, terminal heterocysts at the ends of fragments of trichome, ellipsoidal, oval or subconical, 4.2-8.4μ broad and (4.8-) 6.4-23.5μ long; hormogones present; spores not seen; perennation by the fragmentation of the entire filament remaining dormant inside the persistent sheaths.

Perceived floating with other algae in water bodies as well as in soil cultures of paddy fields.

270. Aulosira pseudoramosa Bharadwaja

Plate-XIX, Fig. 270

Thallus a flat compact system with an uneven surface ordinarily made of unbranched filaments, 9.5-14.7μ broad, up to 2 mm long, generally irregularly bent, densely entangled; sheath thick, hyaline, parallel, stratified with outer surface uneven, firm, 0.75-2.6μ thick, when old becoming deep yellow or golden yellow in colour, hard, thick and brittle; trichome constricted at the cross-walls; cells cylindrical, rarely barrel-shaped, usually slightly longer than broad, sometimes as long as broad, rarely shorter, 6.3- 10.9μ broad; cell-contents blue-green, and fintely granular; heterocysts absent in young trichomes, intercalary, occurring single, rarely in pairs, terminal ones present, as broad as the trichomes, 6.3-10.5μ broad and 6.3-18.9μ long; spore not seen; hormogones one, two or three cells lying dormant in the thick yellowish brown sheath, the filament often breaking into bits or pieces, enclosing one or more hormogones, on germination secreting a thin hyaline sheath, often germinating in situ by piercing through the sheath simulating branching in Tolypothrix.

Perceived in algal collection from water bodies of paddy fields.

Family - Scytonemataceae Rabenhorst

Genus - **Plectonema** Thuret

Trichomes variously bent, with a thin, firm sheath; false branched, branches single or geminate; heterocysts absent; hormogones present; spores not known.

271. Plectonema gracillimum (Zopf) Hansgirg

Plate-XIX, Fig.271

Thallus thin membranous, often later expanded, pale blue, yellowish or grey-green, mostly slimy; filaments 2-4µ broad; false branches single or geminate; cells 1-3 times longer than broad, dull blue-green, nearly colourless; sheath thin, mostly colourless.

Perceived along the edges of paddy fields below water.

272. Plectonema hansgirgi Schmidle

Plate-XIX, Fig. 272

Thallus flat, expanded, up to 20 cm in diam.; filaments densely intricate, 6µ broad, very sparsely false branched; sheath thin, colourless, very much conspicuous, not coloured blue by chlor-zinc-iodide; false branches single, diverging at an acute angle; cells somewhat longer than broad, hormogones small, few-celled, straïght; end cell broadly rounded, often a little less broad.

Perceived aligned to wet paddy leaves.

273. Plectonema nostocorum Bornet ex Gomont

Plate-XX, Fig. 273

Filaments nearly straight or flexuous; false branches sparse, single or geminate; sheath thin, colourless, not coloured by chlo-zinc-iodide; cells cylindrical, sometimes constricted at the cross-walls, septa not granulated, (0.7-) 1-1.5µ broad and 2-2.5 (-3)µ long, pale blue-green; end cell rounded.

Perceived on wet soil surfaces as well as in soil cultures.

274. Plectonema puteale (Kirchner) Hansgirg

Plate-XX, Fig. 274

Filaments in small floccose, pale blue or yellow-green, seldom nearly colourless masses, 3-5 seldom up to 8µ broad, false branches mostly short; cells 2 - 4 (-4.5) µ broad, slightly constricted at the cross-walls, 1- 4 times longer than broad, dull blue-green, or nearly colourless; sheath thin, mostly colourless, 3-8µthick, seldom yellow to brown, thick and opaque.

In soil cultures.

275. Plectonema tomasinianum (Kutz.) Born. ex Gomont

Plate-XX, Fig.275

Filaments flexuous, densely entangled forming a caespitose more or less expanded, brownish green or rarely dull blue-green thallus, thallus up to 2 cm high, richly or sparsely false branched, mostly geminate, filaments 11-24µ broad; sheath up to 3µ thick, lamellated, at first colourless, later yellowish brown; trichome generally constricted at the cross-walls, 11-22µ broad; cells ½ - 1/6 as long as broad, 3-9µ long, seldom longer, blue-green, sometimes granulated at the cross-walls; end cells rounded.

Perceived floating in water bodies of paddy fields.

Genus - **Camptylonemopsis** Desikachary

Filaments bent more or less like a crescent, with a median heterocyst and with two ends of filaments growing upwards, simple or rarely false-branched; true branches absent; trichome forming a single row of cells; heterocysts intercalary or at the base of a false-branch; hormogones present; homocysts not known; spores formed in series.

276. Camptylonemopsis iyengarii Desikachary

Plate-XX, Fig. 276

Filaments epiphytic on Sirogonium filaments, crescent-shaped with a flat attached middle region and lateral erect portions, 5.2-6.6µ broad; sheath thin, hyaline, homogeneous; false-branching not seen; trichome torulose (2.6-) 3.9-5.2µ broad; apex of the trichome slightly broader, rounded; cells spherical, barrel-shaped or sometimes cylindrical, 3.9-5.2µ broad in the middle region and near the apex, 2.6-3.9µ broad and 5.2-10.5µ long in the rest of the trichome; heterocysts intercalary, median heterocysts nearly spherical (3.9-) 5.2-6.6µ broad, in other portions 3.9-5.2µ broad and up to 15.7µ long; hormogones present; spores not seen.

Perceived aligned on wet paddy leaves as well as in soil cultures.

Genus - **Scytonematopsis** Kisselewa

Filaments branched, branching false, gradually attenuated towards the tip, growth by intercalary division; sheath present; heterocysts and spores present.

277. Scytonematopsis woronichinni Kiss.

Plate-XX, Fig.277

Filaments 5-8µ broad, interlacing or sometimes in bundles; the branching invariably false, single or geminate; trichome in the

main filament 5-6.5μ broad, cells mostly quadratic or may be shorter or longer than their breadth and distinctly constricted at the transverse walls, when old almost torulose; cells of the branches 4-6μ broad, as long as broad or up to 12μ long and slightly constricted at the transverse walls, apical cell somewhat conical with an obtuse tip and homogeneous contents; trichomes very slightly and gradually attenuated at the free ends; sheath firm, uniformly thick, unstraited and hyaline, in old filaments yellowish or yellowish brown, not coloured blue or violet with chlor-zinc-iodide; hormogonia formed profusely of varying length, few to many celled; akinetes formed in old cultures, at the terminal ends of the filaments which almost always get displaced during growth as a lateral branch, in chains interspersed with biconcave discs or may be contiguous, oval, oblong or ellipsoidal, 7-10.5μ broad and 10-19μ long with a smooth and firm wall up to 2μ thick.
Perceived floating in water bodies of paddy fields as well as in cultures.

Genus - **Scytonema** Ag.

Filaments false branched, false branches single or geminate, formed laterally generally in between heterocysts; trichomes single in each sheath, straight; hormogones terminal. solitary; pseudo-hormogonia present; spores known only in a few species, spherical or ovate, exospore thin and smooth.

278. Scytonema cincinnatum Thuret ex Born. et Flah.

Plate-XX, Fig.278

Stratum caespitose, intricate, woolly, olivaceous, to brownish green; filaments 16-36μ mostly 18-30μ broad, 3 cm or more long, crisp; false branches mostly geminate, sometimes very sparsely false-branched; sheath firm, membranaceous, hyaline, rarely brownish; trichome 14-30µ broad, blue-green, or olive or brownish violet, distinctly or very tittle constricted at the cross-walls; cells ½ as short as broad, in old filaments nearly as long as broad; heterocysts depressed or quadrate, short cylindrical, or elliptical, single or many.

279. Scytonema fremyii nom. Nov.

Plate-XX, Fig. 279

Thallus on soil, expanded, greenish black; filaments long, about 9µ broad, sparsely false branched; false branches geminate, free at the base, long; sheath yellowish at the apices, hyaline, homogeneous, not lamellate, not dilated, and 5µ thick; trichome constricted at the cross-walls; cells 2-4 times longer than broad, rarely shorter than long; heterocysts rectangular, 4-5 times longer than broad.

Perceived on wet soil surfaces as well as in soil cultures.

280. Scytonema hofmanni Ag. ex Born. et Flah.

Plate-XX, Fig.280

Stratum cushion-like, broadly expanded, 1-3 mm high, blackish blue-green, sometimes impregnated with calcium carbonate, amethyst green or bluish-grey; filaments 7-12µ broad, rarely up to 15µ broad, aggregated in vertical fascicles; false branches aggregated; sheath firm, membranaceous; trichome 5-10µ broad, olive to blue-green, cells unequal in length; heterocysts oblong.

Perceived intermingled with other algal members to form algal mats on wet soil surfaces.

281. Scytonema javanicum (Kutz.) Bornet ex Born. et Flah.

Plate-XX, Fig.281

Thallus cushion-like, 2-4 mm high, deep blue-green to reddish; filament 12-15µ broad, aggregating in erect fascicles; false

branches long flexuous, aggregated; sheath firm, thin, hyaline or becoming yellowish; trichome 9-12μ broad; cells compressed to quadrate; heterocysts subquadrate.
Perceived on much wet soil surfaces as well as in soil cultures.

282. Scytonema millei Barnẹt ex Born. et Flah.

Plate-XX, Fig. 282
Thallus woolly, cushion shape, broadly expanded, 1-5 mm high, dull green or brownish; filaments 15-21μ broad, interwoven, false branches erect; sheath firm brownish, when old with parallel lamellation; cells 10-15μ broad, discoid, when old becoming elongated up to as long as broad; heterocysts mostly discoid, somewhat broader than the trichome.
In soil cultures.

283. Scytonema ocellatum Lyngbye ex Born. et Flah. Var. **purpureum** Gardner

Plate-XX, Fig. 283
Filaments forming a pulvinate or floccose stratum, 1-2 mm thick, 1-1.5 mm long, 14-18μ diam.; cells homogeneous, purplish or pale violet, subquadrate to twice as long as broad, otherwise as in the species.

284. Scytonema pseudohofmanni Bharadwaja

Plate-XX, Fig. 284.
Thallus brownish green, expanded with short tufted growth; filaments 8.4 - 12.6 (-13.6)μ broad, irregularly curved and densely entangled, false-branches few, single and geminate; sheath firm, 1.5 - 2.6 (-3.6)μ thick, pigmented, with rather indistinct parallel stratification; trichome 5.2 - 7.3μ broad, in old filaments narrowed down to 4.2μ septa distinct; cell s usually cylindrical, sometimes quadratic, flattened and barrel-shaped; heterocysts single, sometimes in pairs, usually broader

than the trichome, 5.2 - 8.4μ broad, almost quadratic or ellipsoidal, when deformed spherical or transversely ellipsoidal, 5.2 - 12.6μ long.

Perceived on the wet edges of water bodies in paddy fields.

285. Scytonema schmidlei J. De.Toni

Plate-XX, Fig.285

Thallus attached to tree trunks or leaves, irregularly bluish or violet, rarely greenish; filaments parallel, 5-6μ broad, sparsely false branched, false branches single; sheath thin, hyaline; trichome 3-4μ broad, bluish-green; cells generally 8μlong to subquadrate; heterocysts rare, yellowish, quadrate, rounded or with rounded angles, as long as broad or 1½ times as long.

In soil cultures.

Genus - **Tolypothrix** Kutzing

Filaments with a generally firm, thin or thick sheath with a single trichome in each sheath; false branched, mostly free, prostrate or erect; false branches single mostly sub-tending a heterocyst, occasionally geminate as in Scytonema; hormogonia formed from the tips; trichome with apical growth, apices often broader with shorter cells; spores known in some species.

286. Tolypothrix bouteillei (Breb. et Desm.) Forti

Plate-XXI, Fig.286

Thallus circular, caespitose, blackish brown, up to 2 mm high; filaments 1 mm long 5-7μ broad, false-branched; false branches short, deciduous; sheath thin, close to the trichome, colourless to golden yellow; trichomes 4-5μ broad, torulose; cells somewhat shorter than broad; heterocysts basal, rarely intercalary, single.

Perceived on wet soil surfaces with other algae as well as in soil cultures.

287. Tolypothrix byssoidea (Berk.) Kirchner

Plate-XXI, Fig.287

Thallus woolly, cushion-like, brownish or blackish; filaments up to 1 mm long, 10-15μ (rarely up to 17μ) in diam., irregularly false branched; false branches short, erect, curved; sheath thin, close to the trichome yellowish to brownish, fragile, tubular, sometimes somewhat ocreaceous; trichome 9-11μ broad, torulose; cells barrel-shaped, ½ - 1/3 times as long as broad; heterocysts basal, rarely intercalary, single or in twos; spores seen once, in series, ellipsoidal, longer than the vegetative cells, yellowish green.

Perceived on moist soil as well as in soil cultures.

288. Tolypothrix nodosa Bharadw.

Plate-XXI, Fig.288

Stratum thick, mucilaginous, blue-green to yellow-brawn; filaments irregularly bent, densely entangled, 5.2-7.3 (-8.4)μ broad; false branches few, geminate and single, single branches by the side of dead cells, one pored heterocysts or dead-cells adjoining such heterocysts; sheath firm, thin, occasionally slightly thickened opposite the septa, hyaline, often bulged out at the point of branching; trichomes constricted at the joints, septa indistinct; cells cylindrical, up to 5 times as long as broad, sometimes quadratic; heterocysts intercalary and terminal (basal), single, ellipsoidal, rarely cylindrical with convex walls, when mature much wider than the trichome, 4.2 - 9.4μ broad and 7.3 - 16.8 (-23)μ long.

Perceived floating in water bodies in paddy fields.

Family - Microchaetaceae Lemmermann

Genus - **Microchaete** Thuret

Trichome single with a distinct sheath, more or less attenuated at the apices, seldom equally broad all through; filaments attached, single or in groups or clusters, seldom with single pseudo-branches false branches; heterocysts basal, seldom intercalary; spores mostly single, seldom in series, away from the heterocysts or near it; hormogones present.

289. Microchaete aequalis (Fremy) Comb. Nov.

Plate-XXI, Fig.289

Thallus expanded, blackish blue, filaments sufficiently elongated, more or less intricate, base erect, 4.5-5μ broad; sheath thin, close, papery, colourless, or rarely yellowish or brownish, coloured blue by chlor-zinc-iodide trichome uniform, 3-3.5μ broad, not constricted at the cross walls; cells 2-4 times as long as broad; spores and hormogones not seen.

Perceived aligned to submerged paddy leaves as well as in soil cultures.

290. Microchaete calothrichoides Hansgirg

Plate-XXI, Fig.290

Filaments 10-16μ broad, seldom up to 20μ broad, single or in tufts, forming a dirty greyish green thallus, straight or curved; sheath thick, lamellated often more or less incrusted, colourless; cells at the base of filaments 6-8μ broad, 1/3 to as long as broad, distinctly constricted at the cross-walls, olive coloured, heterocysts mostly basal, nearly oval to long ellipsoidal, 6μ broad, up to 8μ long.

Perceived aligned to paddy in submerged parts as well as in soil cultures.

291. Microchaete loktakensis Bruhl et Biswas

Plate-XXI, Fig.291

Filaments aggregated, long, uniformly cylindrical, a little broader at the base (2μ), flexuous, briefly decumbent at the base, 12-14μ broad, above about 8μ broad, at the apical portions about 6μbroad, sheath 1.5 - 2.3μ broad, with lamellae, colourless, to pale blue green; trichome below 6 - 8μ broad, with cells 1 – 1½ as long as broad, and 6-10μ long, above about 8μbroad and cylindrical cells, 1½ - 2 as long as broad and 8-16μ long; heterocysts generally basal, spherical to ellipsoidal, 6 - 9μ diam., intercalary heterocysts absent or rare, 16-24μ long, 8μ broad; cells contents granular.

Perceived free floating in water bodies in paddy fields.

292. Microchaete tenera Thuret ex Born. et Flah.

Plate-XXI, Fig. 292

Filaments up to 1 mm long, single or in small stellate clusters, 6-7(-.5)μ broad, slightly bent; prostrate at the base; sheath thin, close to the trichome, hyaline, unlamellated; trichome 5μ broad, blue-green; cells at the base twice as long as broad, at the apices as long as broad; heterocysts basal and intercalary, nearly spherical or cylindrical, 6μ broad, 6 - 8.5μ long; spores single or in series, basal or intercalary, cylindrical with brownish wall, 6 - 7.5μ broad, 13-17μ long.

Perceived on wet soil surfaces along with other alga as well as in soil cultures.

293. Microchaete tenera Thuret ex Born. et Flah. Var. **tenuis** Bharadwaja·

Plate-XXI, Fig. 293

Filaments up to 250μ long, up to 4.2μ broad; trichome 3.1 - 3.6μ broad; heterocysts not cylindrical, 3.6 - 4.2μ diam. Perceived aligned to paddy leaves in water filled fields.

294. Microchaete uberrima Carter, N.

Plate-XXI, Fig. 294

Filaments elongate, up to 5 mm long, absolutely cylindrical, decumbent, at their bases stellately arranged, (9-) 16-18µ broad; sheath firm, brown; trichome (6.5-) 10-14µ broad, not attenuated; cells generally subquadrate; heterocysts basal and intercalary, cylindrical spherical and slightly adpressed; spores in long series, olivaceous, quadrate to subcylindrical.

Perceived intermingled with other alga in water bodies of paddy fields.

295. Microchaete uberrima Carter, N. f. **minor** Carter

Plate-XXI, Fig. 295

Filaments stellately arranged at the base, 9-11µ broad, brown; sheath firm; trichome 6.5 - 8.0µ broad, not attenuated; cells 4-6µ long, heterocysts spherical, slightly adpressed, 6.6 - 8.0µ broad; spores in series.

In soil cultures.

Genus - **Fortiea** De Toni, J.

Filaments unbranched, trichomes with a firm sheath, attenuated at the base, broadened at the apices, with a basal and intercalary heterocysts, generally attached by the basal end; hormogones and spores present.

296. Fortiea incerta Skuja

Plate-XXI, Fig. 296

Filaments many in caespitose clusters, rarely solitary more or less erect, unbranched, attached to the substratum by the base, base and apices clearly dilated, 160 - 650µ or more long, 7 - 9.5µ broad below, and up to 14µ broad above; sheath thin, about 1µ thick, unlamellated, firm, colourless; trichome in basal part 5.5

- 6μ broad, and at the apex 11μ broad, clearly constrictm9 at the cross-walls; cells as long as or twice as long as broad, 6 - 16μ long, above ¾ - ½ as long as broad, contents olivaceous, or pale blue-green; heterocysts basal, rarely intercalary, single or two together, globose or cylindrical and discoid, yellowish 8-10μ broad and 8-15μ long; spores not seen; hormogones short or unicellular formed from the apices.

Perceived aligned to submerged paddy plants.

Genus - **Calothrix** Ag.

Filaments single or in small bundles, caespitose, tomentose, pulvinate, or penicillats; filaments arranged more or less parallel, mostly erect, unbranched or seldom false-branched; sheath mostly firm, sometimes seen only at the base; heterocysts mostly basal, seldom intercalary; spores when formed single or in series, next to the basal heterocyst.

297. Calothrix braunii (A.Br.) Bornet et Flahault

Plate-XXI, Fig. 297

Thallus caespitose, blue green or brownish; filaments straight, parallel, 500μ long, 9 - 10μ broad, swollen at the base, slightly bent; sheath thin, close to the trichome, colourless, trichome 6-7μ broad, ending in a long hair, constricted at the cross-walls; cells somewhat shorter than broad; heterocysts basal, hemispherical.

Perceived on wet soil surfaces as well as in soil cultures.

298. Calothrix brevissima West, G.S.

Plate-XXII, Fig. 298

Filaments epiphytic, many together, very short, 53-94μ long, 5-7μ broad, not at all or very little attenuated; sheath firm, close to the trichome, thin, nearly cylindrical colourless; trichomes

very short, 30-62μ long, little attenuated, with rounded end cells, not constricted at the cross-walls olive green; cells at the base nearly as long as broad; heterocysts basal, single seldom many, rounded, hemispherical or subspherical.
Perceived alighed with free floating mass of algae.

299. Calothrix brevissima West, G.S. Var. **moniliforme** Ghose orth. mute Geitler

Plate-XXIII, Fig.299
Filaments gregarious, 5-7μ broad, not at all or very little attenuated; sheath indistinct, narrow, colourless, trichomes pale olive green, very short, 30-62μ long, with rounded end cells, moniliform; cells at the base as long as broad or shorter, globose or depressed globose, 4-5μ broad.
Perceived free floating in water bodies as well as in cultures.

300. Calothrix castellii (Massal.) Born. et Flah.

Plate-XXIII, Fig.300
Thallus spongy, cushion shaped, widely expanded, dull blue green, surface pubescent by projecting ends of filaments; filaments bent, erect, densely aggregated, 12-13μ broad, swollen at the base and prostrate, 4-8 mm long; sheath thin close to the trichome, firm, uniform, hyaline, or yellowish; trichome 8-10μ broad, attenuated into a long hair; cells ½- as long as broad; heterocysts basal.
In soil cultures.

301. Calothrix clavata West, G.S.

Plate-XXIII, Fig. 301
Filaments single, or a few together, up to 100μ long, straight or slightly bent, prominently swollen at the base, 7μ broad; sheath close to the trichome, very thin, colourless; trichomes slightly constricted at the cross walls, 5-5.5μ broad at the base, in the

middle 2-5μ broad; cells discoid at the base and 2-3 times as long as broad above; heterocysts basal, single hemispherical. Perceived free floating in water bodies of paddy fields.

302. Calothrix epiphytica W. et G.S. West

Plate-XXIII, Fig.302

Filaments single or in groups, up to 250μ long, seldom up to 350μ long; at the base 5-7.5μ broad, at the apex gradually attenuated; sheath somewhat thick, colourless; tricome at the base, 3.5-4μ broad tapering to a fine hair cells somewhat shorter than broad below and longer than broad above; heterocysts basal, single, small.

Perceived aligned to submerged paddy parts as well as in soil cultures.

303. Calothrix elenkinii Kossinskaja

Plate-XXIII, Fig.303

Filaments 80-250μ long, united in tufts, bent at the base, interlaced with each other, swollen at the base, and 6-9μ broad, in the middle 4.5-6μ broad; sheath close to the trichome, thin, not lamellated, colourless, open at the ends; trichomes blue or olive green, at the base 5-1μ broad, in the middle 3.5 - 4.5μ broad, not constricted at the base, apical hair not formed; cells quadrate or somewhat shorter than; heterocysts basal, single 4.5 - 7μ broad.

Perceived aligned to submerged paddy leaves.

304. Calothrix javanica de Wilde.

Plate-XXIII, Fig. 304

Filaments single, in the mucilage of other algae, with a distinct colourless, unlamallatcd, sheath, gradually attenuated to a pointed apex, at the base not swollen; trichomes 4-6μ broad,

ending in a thin hair; heterocysts basal, 4-5.5μ broad; spores single or two together, about 4μ broad and 6-10μ long.
Perceived in algal mat collections from paddy fields.

305. Calothrix membranacea Schmidle

Plate-XXIII, Fig. 305
Thallus papery, blue green; filaments long, horizontal, mostly curved, irregularly floccose, rarely branched, gradually attenuated; sheath thin, hyaline, not lamellated; trichomes torulose, only seldom trichome produced into a hair; cells half as long as broad to subquadrate, in the apex and the hair more or less hyaline and elongate; hormogones with a few cells, formed in series; heterocysts present, basal.
Perceived intermingled with other alga in soils with more water contents as well as in soil cultures.

306. Calothrix marchica Lemmermann

Plate-XXIII, Fig. 306
Filaments straight, or slightly bent, single, at the base 5-6μ broad, with a close thin colourless sheath; sheath not coloured blue by chlor-zinc-iodide; trichome blue green, gradually attenaated into a hair, distinctly constricted at the cross-walls, at the base 4-5.5μ broad; cells nearly as long as broad, or , ½ to ¼ as long as broad, end cell conical, somewhat pointed; heterocyst single, basal nearly spherical or hemispherical, 4-5.5μ broad.
In soil cultures.

307. Calothrix marchica Lemmermann Var. **crassa** Rao, C.B.

Plate-XXIII, Fig. 307
Filaments in groups, irregularly bent and closely entangled, 9.6-14.4μ broad, up to 450μ long; sheath thin, firm, yellowish, or hyaline; trichomes 8.4-12.8μ broad, constricted at the

septa, ends tapering but without a hair, end cell conical with a rounded apex, sometimes pointed cells quadratic, as well as shorter or longer than broader, 2-3.3µ long, at the apices up to 4.8µ long; heterocysts single, basal, spherical or subspherical, 8.2-12.5µ broad and up to 5µ long.
Perceived on the edges of water bodies in paddy fields.

308. Calothrix marchica Lemmermann Var. **intermedia** Rao, C.B.

Plate-XXIII, Fig. 308
Filaments epiphytic on other algae, placed singly or in groups of two or three, 6-8µ broad and up to 350µ long, with slight attenuation, without a terminal hair; sheath thin, firm, hyaline; trichomes 5.8-7.8µ broad, constricted at the cross walls; cells quadratic, shorter or longer than broad, 3.2-8µ long, end cells rounded; heterocysts single, basal, usually spherical, 5.6-7.8µ broad.
Perceived free floating in water bodies as well as in soil cultures.

Genus - **Dichothrix** Zanardini

Thallus caespitose, penicillate or cushion like (pulvinate); filaments subdichotomously false branched, at the base of false branches often many trichomes in a common sheath; trichomes generally ending in a hair; heterocysts mostly basal, sometimes intercalary.

309. Dichothrix baueriana (Grun.) Born. et Flah

Plate-XXIII, Fig. 309
Thallus caespitose, penicillate, often very much expanded, and up to 1 cm high, green or brown; filaments flexuous, ultimate branches 15-21µ broad; sheath close to the trichome,

gelatinous, soft, uniform, hyaline or yellowish; trichome 5-7.5μ rarely 9μ broad, pale blue-green, gradually terminating in a hair, constricted at the cross walls, cells as long or ½ as long as broad; heterocysts basal subspherical or hemispherical.

In soil cultures.

Genus - **Rivularia** (Roth) Ag.

Trichomes unbranched, more or less irregularly false branched; filaments more or less radial or parallel in a hemispherical or spherical mucilaginous colony, hollow or solid; sheath more or less gelatinising; trichomes ending in a hair, often with distinct trichothallic growth; heterocysts basal or intercalary; often false branching at the base; hormogonia single or in series gradually progressing towards the base from the meristematic zone; spores absent.

310. Rivularia beccariana (De Not.) Born. et Flah.

Plate-XXIII, Fig. 310

Thallus hemispherical, pulvinate, sometimes confluent, olivaceous, green, hard, 1 mm thick, seldom up to 3.5 mm high; filaments 7-9μ broad, very closely pressed together; sheath thin, indistinctly lamellated, colorless or yellow-grown; trichome 3-7μ broad, olive-green, produced into a long very much bent hair; cells at the base longer than broad, above as long as broad.

Perceived aligned to soil surfaces below water bodies as well as in soil cultures.

311. Rivularia bullata (Poir) Berk. ex Born. et Flah.

Plate-XXIII, Fig. 311

Thallus expanded, vesicular, lobed, bullose-confluent, up to 6 cm. diam., blue-green; filaments not easily seen owing to being closely adpressed; sheath distinct, hyaline, seldom yellowish, diffluent or gelatinized and lost in the upper part; trichome 5-8 rarely 10μ broad, attenuated into a hair, slightly constricted at the cross-walls; cells at the base as long as broad, and above up to 1/3 as long as broad.

Perceived on wet soil surface as well as in soil cultures.

Genus - **Gloeotrichia** Ag.

Thallus spherical or hemispheridal, solid, sometimes when old inflated and hollow; filaments fadial more or less parallel, often with false branches; sheath at the base firm only gelatinizing on the outside, soft to mostly diffluent; trichomes with a distinct trichothallic growth; heterocysts basal; spores at the base of the trichome, single or a few next to the heterocyst; hormogones present.

312. Gloeothrichia indica Schmidle

Plate-XXIII, Fig. 312

Thallus spherical, soft, 1-2 mm diam., hollow (or solid), filaments radiating 200-300μ long; trichomes made of 2-3 cells, cells barrel-shaped and about 8μ broad, and a long narrow hair with cylindrical cells; spores nearly cylindrical, 60-70μ long, without sheath 18-20μ broad, with sheath 20-26μ broad, sheath at first hyaline later yellowish brown and close to the trichome.

Perceived in water bodies of paddy fields.

313. Gloeothrichia intermedia (Lemm.) Geitler Var. **kanwaensis** Rao, C.B.

Plate-XXIII, Fig. 3l3

Thallus spherical, soft, 1-10 mm diam., filaments with a thin, hyaline and closely adpressed sheath; trichomes with constrictions at the joints at the base, 10-13µ broad, higher up 4-6µ broad; cells at the base of the trichome invariably flattened and barrel-shaped, higher up almost quadratic or cylindrical, in the hair long cylindrical board, at the base 2.5 - 4.5µ long, higher up 5-16µ long with pale blue-green contents; heterocysts spherical or subspherical, usually single, rarely in pairs, 8.6-11.5µ long; spores cylindrical with smooth outer walls, with sheath 14-17µ borad, without sheath 12 - 15µ broad, 55 - 66µ long, sheath 0.8 - 1µ thick.

Perceived in water bodies of paddy fields.

314. Gloeothrichia natans Rabenhorst ex Born. et Flah.

Plate-XXIII, Fig.314

Thallus spherical, up to 10 cm broad, soft, bullate, hollow, blackish olive green to brown, filaments loosely arranged; trichome 7-9µ broad, olivaceous tenuated into a long hair; cells at the base barrel-shaped, as long as broader somewhat shorter, higher up to 4 times as long as broad; heterocysts basal, more or less spherical, 6-12µ broad; spores cylindrical, straight or bent, without sheath 10-18µ broad and 40-250µ long, with sheath up to about 40µ broad, saccate, transversely constricted, hyaline, or brownish.

In soil cultures.

Order	-	Stigonematales Geitler
Family	-	Nostochopsidaceae Geitler
Genus	-	**Nostochopsis** Wood em. Geitler

Thallus attached at first, later free-floating with erect filaments with soft diffluent sheath, mucilaginous, more or less hemispherical, at first solid, later hollow, torn and expanded; trichome a single or two rows of branches, branches of two types, one long and many celled, the other one of limited growth with a heterocyst at the end (pedicellate heterocyst), pedicels made of one to three cells; heterocysts intercalary or terminal or pedicellate, or lateral and sessile; hormogones present; spores not known.

315. Nostochopsis lobatus Wood em. Geitler

Plate-XXIII, fig. 315

Thallus spherical or Sub-spherical, more or less irregularly lobed, solid or hollow up to 3.5 cm diam., seldom very big, blue-green or olive-green or yellowish green; trichomes radically arranged seemingly straight, in the innermost parts bent irregularly or in a zig-zag manner, lower part of the upper region richly branched, in the older parts branches scattered; branches later, broadly stretching upwards by a short bend and with its end being parallel to the main filament; cells in the branching zone barrel-shaped 4-5.5μ broad and just before formation up to 9μbroad, mostly up to 8μ and in older regions up to 18μ long, trichome end tapering or slightly pointed; heterocysts mostly lateral (seesile) or terminal (pedicellate) on 2-3 or 4 celled lateral branch, more or less spherical or ellipsoidal, mostly 6-9μ broad, seldom intercalary or basal (when young) mucilage homogeneous, colourless or yellowish or brownish, sheath present in the branching zone, thin and firm, seemingly colourless or coloured yellowish to brownish, diffluent at the apices of branches, but persistent at the base; hormogones on special branches, 3-3.5μ broad, and 8-20 cells length, cells short barrel-shaped.

Perceived in algal collections from water bodies of paddy fields.

Family - Stigonemaceae Kirchner

Genus - **Hapalosiphon** Nag.

Thallus caespitose, floccose, thin, aquatic; filaments free, not coalescing laterally; cells in one or two rows, sheath present, continuously branched; branches irregularly lateral, true, often arising only on one side of the filaments false branches present branches erect from the primary prostrate filaments, erect branches as broad as and similar to the main filament; heterocysts intercalary- only occasionally lateral; hormogones formed mostly from the side branches; spores present.

316. Hapalosiphon baronii W. et G.S. West

Plate-XXIII, Fig. 316

Thallus caespitose; filaments densely aggregated, richly branched 7.5-10μ broad; sheath broad, colourless; cells of the lateral branches quadrate, rounded to subspherical; 3.5-5μ diam., lateral branches short; heterocysts subquadrate.

Perceived free floating in water bodies of paddy field.

317. Hapalosiphon hansgirgi (Schmidle) Forti

Plate-XXIV, Fig. 317

Thallus papyraceous, mostly single layered, pale blue green; filaments irregularly intricate and curved, sometime parallel, cylindrical; filaments 8μ diam., narrowing down to about 5μ occasionally not branched, rarely false branched, branches short, without sheath; sheath hyaline, mucilaginous; trichome blue green, 6-8μ diam., cells isodiametric or half as long as broad, at first cylindrical, later torulose, contents granular.

Perceived in algal collections from water bodies of paddy fields.

318. Hapalosiphon intricatus W. et G.S. West.

Plate-XXIV, Fig.318

Thallus caespitose, blue-green, small, filamently densely interwoven, sparsely branched, 1 4-7μ broad; sheath close to the trichome, colourless often indistinct, cells spherical, to cylindrical, 1½ -3 times as long as broad; lateral branches as thick as the main filament; heterocysts intercalary, subquadrate to cylindrical, 3.8 - 5.5μ broad; spores spherical to ellipsoidal, seldom nearly cylindrical.

Perceived aligned to submerged paddy leaves as well as in cultures.

319. Hapalosiphon welwitschii W. et G.S. West

Plate-XXIV, Fig. 319

Filaments single among other algae, somewhat flexuous, 5.5-7.5μ broad; sheath very close, hardly visible, colourless; cells subspherical or elongate, as long as broad or longer; lateral branches short, as broad as the main filament or narrower, 3.5-5.7μ broad, slightly attenuated at the ends; cells of the branches ½ - 3 times as long as broad; heterocysts rare, intercalary, quadrate rounded or long cylindrical, 6μbroad, 6-8μ long; spores sub-spherical or oblong, 5μ broad, 1-2 times as long as broad.

Perceived on wet soil surfaces as well as in soil cultures.

Genus - **Westiellopsis** Janet

Thallus filamentous with true branching filaments of two kinds, primary filaments slightly thicker and more or less creeping, secondary filaments generally thinner and growing erect, filaments without a sheath and consisting of a single row of cells; heterocyst intercalary; the dilated terminal portions of secondary branches by profuse transverse and longitudinal divisions forming chains of rounded cells (Pseudohormocysts), the contents escaping as gonidia (endospores) and develop as into new plants.

320. Westiellopsis prolifica Janet

Plate-XXIV, Fig. 320

Main filaments torulose, with short barrel-shaped cells, 8-12μ broad as long as broad, or slightly longer, branch filaments thinner and elongate, not constricted at the cross-walls, with elongate cylindrical cells, 4-6μ broad; heterocysts oblong-cylindrical, 5.5-6μ broad and 10.5 - 22μ long; gonidia formed singly in each cell of the pseudohormocysts, 8.5-9μ diam. Perceived on wet soil surfaces, as well as in soil cultures.

Genus - **Fischerella** (Born. & Flah) Gomont.

Thallus creeping, most of it or only portions with only portions of filaments with cells in many rows, seldom filaments with a single row of cells, erect branches only on one side; lateral branches with long and narrow cells, main filaments with large spherical cells; sheath in young lateral branches thin and close to the trichome, sheath in the older filaments thicker; cells of old filaments often with Gloeocapsa like sheath; heterocysts intercalary or lateral; hormogones from the ends of side branches; spores known in some species.

321. Fischerella muscicola (Thuret) Gom.

Plate-XXIV, Fig. 321

Main filaments creeping, torulose, flexuous inter-woven, forming a thick blackish brown thallus, filaments about 10μ broad, mostly with two rows of cells; cells sub-quadrate, subspherical, 7.5μ diam; sheath of the main filament, close to the trichome, thin, brown; lateral branches erect, more or less straight, 6μ broad, with a thin colourless or yellowish sheath, with false branches; hormogones present; cells of the lateral branches subquadrate or shorter than broad, blue green; heterocysts subspherical, mostly smaller than the vegetative

cells; hormogones long, sometimes club-shaped, 4μ broad, 100 μ long.
In soil cultures.

Genus - **Stigonema** Ag.

Thallus of free irregularly laterally branched, variously curved filaments in older parts with two to many rows of cells; sometimes with apical growth and division; lateral branches like the prostrate parent filaments and as many rows of cells as the main filaments; sheath when young close to the trichome, in older ones broader; old filaments with Gloeocapsa-like sheath and cell grouping; heterocysts intercalary or lateral; hormogonia formed at the end of young branches, two to a few celled, seldom many celled.

322. Stigonema hormoides (Kutz.) Born. at Flah.

Plate-XXIV, Fig. 322

Filaments prostrate, densely intertwined, long, forming a thin tomentose, blackish brown thallus; filaments 7-15μ broad, irregularly and sparsely branches, lateral branches, erect, nearly as thick as the main filament, curved, sheath thick, colourless or yellow to yellowish brown; cells subspherical, mostly single rowed, occasionally in two rows; heterocysts sparse, intercalary, or lateral; Chroococcoid conidia frequently present;
Perceived on moist soil surfaces as wall as in soil cultures.

323. Stigonema lavardei Fremy Var. **indica** Fremy

Plate-XXIV, Fig. 323

Cells of filaments and branches somewhat longer; heterocysts lateral, many; hormogoniferous branches less broad and somewhat longer.
Perceived in algal mat collections from wet soil surfaces of paddy fields.

Division - Chlorcphyta
Class - Chlorphyceae
Order - Chlamydamonadaceae.
Genus - **Chlamydomonas**. Ehrenberg.

1. Chlamydomonas globosa Snow.

Plate - XXV, Fig. l.
Cells broadly ovoid to globose, surrounded by a hyaline gelatinous sheath; chlaroplast a dense parietal cup with a basal pyrenoid, pigment spot lens-shaped; cells 6-7μ broad, 6-10μ long.
Perceived in algal collection s from water bodies of paddy fields as well as in soil cultures.

2. Chlamydomonas lismorensis, Playton.

Plate - xxv, Fig.2.
Cells smaller in size, 7.2 - 9.6μ long, and 4.8 - 6.4μ broad; chloroplast cup shaped with single basal pyrenoid, flagella as long as the cells.
Perceived in algal collections from water bodies of paddy fields.

3. Chlamydomonas mucicola Schmidle.

Plate - XXV, Fig. 3.
Cells narrowly elliptic to ovoid, attenuated to a blunt point anteriorly, broadly rounded posteriorly, chloroplast cup shaped with a large pyrenoid; cells 5.6-6.4μ broad, 8-10μ long.
Perceived in soil cultures.

4. Chlamydomonas reinhardi Dangeard

Plate - XXV, Fig. 4.

Cells ovoid; length 11.2 - 14.4μ and breadth about 10μ; chloroplast cup-shaped lining the wall; phrenoid at the lower side; flagella slightly longer than the length of the cell.
Perceived in soil cultures.

5. **Chlamydomonas sphaginicola** Fritsch. et. Takeda.

Plate - XXV, Fig. 5.
Cells broadly ovoid to globose, with two prominent papillae at the anterior end; chloroplast a parietal sheet, with 3-4 pyrenoids; pignent spot prominent, anterior to median; cells 15-19μ broad, 18-22μ long.
Perceived in water bodies of paddy fields.

Genus - **Carteria** Diesing.

6. **Carteria klebsii** (Dang.) Dill.

Plate-XXV, Fig. 6
Cells ellipsoid to ovoid, anterior end narrower, sharply rounded and with a papilla; chloroplast a massive parietal cup, with a pyrenoid; pigment spot not very prominent; cells 5-8μ broad, 8-10μ long.
Perceived in soil cultures.

Order - Chlorococcales
Family - Chlorococcaeeae
Genus - **Characium** A. Braun.

7. **Characium ambiguum.** Hermann ex Rabenhorst

Plate-XXV, Fig. 7
Cells 5.6-8.4μ broad and 39-56μ long.
Perceived in algal mat collections from wet paddy field soils.

Order - Chlorococcales
Family - Chlorococcaceae
Genus - **Chlorococcum** Meneghini emend. Starr.

8. **Chlorococcum humicolo** (Naegeli) Rabenhorst

Plate-XXV, Fig.8

Cells spherical, solitary or a number of cells crowded together to form a stratum; chlaroplast, a hollow sphere with a lateral notch and a single pyrenoid; cells 2-2.5µ in diameter.

Perceived in algal mat collections from wet soil surfaces as well as in soil cultures.

9. **Chlorococcum infusionum** (Schrank) Meneghini

Plate - XXV, Fig.9.

Cells spherical, in flat irregular colonies; colony upto 96µ broad; cells without sheath 12.8µ and with sheath 20.8µ in diameter, with a single pyrenoid.

Perceived in algal collections from wet soil surfaces as well as in soil cultures.

10. **Chlorococcum vitiosum** Printz.

Plate - XXV, Fig.10.

Cells usually spherical with a thin cell wall which is usually smooth, but sometimes crenulate on the outside locally or allover the surface; chloroplast hemispherical, reticulate, without a pyrenoid, nucleus central; cells 8-22µ in diameter.

Perceived aligned on the submerged leaves of paddy.

Genus - **Trebouxia.**

11. **Trebouxia humicolo** (Treboux) West. et Fritsch.

Plate - XXV, Fig.11.

Cells spherical, solitary, or in regular colonies, 7.2-12.8µ in diam., sheath thin colourless; chloroplast central with lobed margines.

Perceived in algal collections from water bodies of paddy field.

Family - Chlorellaceae.

Genus - **Chlorella.**

12. Chlorella conglomerata (Artari) Oltamanns.

Plate - XXV, Fig.16.
Colonies with many cells; cells spherical, chlorcplast hollow, spherical with a pyrenoid.
In soil cultures.

13. Chlorella vulgaris Beijerinck.

Plate - XXVI, Fig.13.
Cells small, solitary, spherical to broadly ellipsoid; with thin cell membrane 6.4 - 8.0μ in diameter; chloroplast cup shaped and with a single pyrenoid.
Perceived in algal mat collections from wet soil surfaces of paddy field as well as in soil cultures.

Genus - **Tetraedron** Kuetzing.

14. Tetraedron guadratum (Reinsch) Hansgirg. f. **minus** (Reinsch.) De Toni

Plate - XXVI, Fig.14.
Cells solitary, flat, quadrate, 9.6-12.0μ in diam., angles of cells with spines; sides of cells convex, cell membrane thick; length of spine 1.6-2.4μ.
In soil cultures.

15. Tetraedron müticum (A. Braun) Hansgirg.

Plate - XXVI, Fig. 15.
Cells single, triangular, pale-green; angles rounded and sides slightly concave, 20.8 - 30.0μ in diameter.
Perceived in algal collection from water bodies of paddy fields.

16. Tetraedron octaedricum v. **spinosum** (Reinsch.) W.et. G.S. West.

Plate - XXVI, Fig.16.

Cells solitary, octagonal; angles rounded, each with a spine; cell 11.2μ in diameter, length of spine 1.6μ.
Perceived free floating in water bodies of paddy field.

Family - Oocystaoeae.
Genus - **Trochiscia** Kuetz.

17. Trochiscia aciculifera (Lagerb.) Hansgirg.
Philipose, M.T., 1967, p. 99, Fig. 22.
Plate - XXVI, Fig. 17.
Cells solitary, spherical, sheath thick and covered by numerous sharply pointed spines; cells 12.8μ in diameter; spines 1.6μ long.
Perceived in algal collections from small water bodies in paddy field.

18. Trochiscia granulata (Reinsch) Hansgirg.
Plate - XXVII Fig. 18.
Cells spherical, 19.2μ in diam, covered with small spines all over the surface.
In soil cultures.

19. Trochiscia reticularis (Reinsch.) Hansgirg.
Plate - XXVI, Fig.19.
Cells solitary, spherical, 21.6 - 24.8μ in diam., cell wall fairly thick forming linear ridges arranged to form polygons.
Perceived in the algal collections form water bodies in paddy field.

Family - Oocystaceae.
Genus - **Oocystis** Naegeli.

20. Oocystis elliptica W. West.
Plate - XXVI, Fig.20.
Cells in groups of 2-4, within the old mother cell wall; colony envelope irregular; cells ellipsoid, with poles of cells broadly

rounded and without polar thickenings; chloroplasts numerous, parietal discs; cells 17-19μ in diameter, 25-27μ long.
Perceived in water bodies of paddy field.

21. Oocystis lacustris. chodat.

Plate-XXVI, Fig.21
Cells solitary or 2-8 within the old mother cell wall; cells broadly elliptic or moniliform, the poles furnished with large nodular thickenings; chloroplast solitary, parietal, usually with a pyrenoid; cells 12-22μ broad, 25-31μ long.
Perceived in algal collections from water bodies of paddy fields.

22. Oocystis nodulosa West et West.

Plate-XXVI, Fig.22
Cells solitary or two within the old mother cell wall, ellipsoid to oblong-ellipsoid with rounded apices bearing a papillate thickening which projects both inward and outward; cells (6-) 12-16μ broad, (12-) 28-32μ long.
Perceived floating along with other algae in water bodies of paddy fields.

Family - Protosiphonaceae
Genus - **Protosiphon.**

23. Protosiphon botryoides (Kutz.) Kelbs.

Plate- XXVI , Fig. 23
Thallus unicellular, upper part broad; about 16-55μ broad, lower part narrow, tapering 14-20μ broad; aplanospores globose 17.5-24.5μ in diameter.
Perceived in algal collections from water bodies of paddy field.

Family - Selenastraceae
Genus - **Ankistrodesmus** Corda.

24. Ankistrodesmus falcatus (Corda) Ralfs.

Plate-XXVI, Fig.24

Cells acicular to spindle-shaped, with ends tapering to acute apices; solitary or in clusters of 2-32; not enclosed in colonial sheaths; chloroplasts solitary, parietal, without a pyrenoid; cells 1.3-3µ in diameter, 35-60µ long.

Perceived free floating in water bodies of paddy

25. Ankistrodesmus falcatus (Corda) Ralfs. V. **tumidus. (West et West) G.S. West.**

Plate-XXVI, Fig.25

Cells solitary, rarely in small aggregates, usually straight, rarely slightly curved, inflated in the middle, apices acute; chloroplast single, parietal with or without a pyrenoid; cells 3-3.8µbroad, 25-75µ long.

Perceived in algal mat collections from wet soils of paddy fields.

Genus - **Selenastrum**. Reinsch.

26. Selenastrum gracile Reinsch.

Plate-XXVI, Fig.26

Colonies of 8-64 sickle-shaped cells in irregular arrangement but with their convex surface apposed; apices of the cells sharpely pointed; chloroplast solitary, parietal with a pyrenoid; cells 2-3.3µ in diameter, 25-40µ long, distance between the apices (6-) 16-20µ.

Perceived in water bodies of paddy fields.

Family - Coclastraceae

Genus - **Scenedesmus** Meyen.

27. Scenedesmus arcuatus (Lemmermann) Limmermann.

Plate-XXVII, Fig.27

Colonies 4-8 celled, curved and with small inter cellular spaces, cell wall smooth without spines, cells 5-8μ broad, 13-16μ long.
In soil cultures.

28. Scenedesmus bijugatus (Turpin) Kutzing Var. **Bicellularis** (Chodat) comb.nov.

Plate - XXVII, Fig. 28.
Colonies 2 cells, solitary, cells ellipsoid, cylindrical cells also not uncommon; colonies arranged in group of 2; cells 7-10μ long and 2-4μ broad.
Perceived in algal collections from water bodies of paddy fields.

29. Scenedesmus bijugatus (Turpin) Kutzing Var. **irregularis** (Wille) Smith.

Plate - XXVII, Fig.29.
Colony composed of a 4 cells arranged irregularly in double series; cells ovate-oblong with rounded ends, 5-8μ in diameter, 14-16μ long.
Perceived in algal mat collections from much wet soil surfaces of paddy fields.

30. Scenedesmus dimorphus. (Turpin) Kuetzing.

Plate - XXVII, Fig.30.
Colonies 4-8 celled with the cells arranged in sub-alternating series, outer cells are lunate and the apices being attenuated; cells 3-8μ broad, 10-17μ long.
Perceived in algal collections from wet soil surfaces as well as in soil cultures.

31. Scenedesmus falcatus Chodat

Plate - XXVII, Fig. 3l.

Colony composed of 4-8 cells arranged in a single or double series; cells fusiform; inner cells straight or curved with sharp apices, outer cells lunate; strongly curved; rarely all the cells in one plane or the interior cells forming a flat plate and the outer cells lunate and at an angle to the interior cells; cells without any teeth or spines; cell wall smooth; cells 3-6μin diameter, 13-16μ long.

Perceived in the collections from water bodies of paddy fields.

32. Scenedesmus longus v. brevipseina. Smith.

Plate - XXVII, Fig.32.

Colony composed of 4 cells arranged in a single series; cells ovoid to naviculoid with rounded or somewhat acute ends; poles of cells with a short spine; cells 3.5-4.5μ in diameter, 10-11μ long.

Perceived in the collections from water bodies of paddy fields.

33. Scenedesmus oblicuus. (Turpin) Kuetzing.

Plate - XXVII, Fig. 33.

Colonies of usually 4 cells, in linear series, cells fusiform with acute apices, outer sides of terminal cells concave; cell wall smooth without spines, cells 3.2-4.0μ broad and 14.4-16.0μ long.

Perceived in algal collections from water bodies of paddy fields.

34. Scenedesmus quadricaude v. **maximus** W. et G.S. West

Plate-XXVII, Fig.34

Colony composed of 4-8 cylindric-ovate cells arranged in a single series; outer cells with long, curved spines; cells 10-15μ in diameter 25-35μ long.

Perceived in algal collections from water bodies of paddy fields.

Order - Ulotrichales
Sub-order - Ulotrichineae
Family - Ulotrichaceae
Genus - **Ulothrix** Kuetzing.

35. Ulothrix oscillarina Kutzing.

Plate-XXVII, Fig. 35

Cells 8-11μ broad, occasionally up to 14μ and ½ - ½ times as long as broad; cell wall thin, easily gelatinising; chloroplast a broad girdle-shaped band, covering more than half the circumference and containing 2 or 3 pyrenoids.

Perceived in algal collections from water bodies of paddy field.

36. Ulothrix subtilissima Rabenhorst

Plate-XXVII, Fig.36

Filaments long and slender, free floating or attached; cells 4-5μ broad and 8-12μ long, occasionally longer; chloroplast extending nearly the entire length of the cell, with one or two pyrenoids.

Perceived in algal collections from water bodies of paddy field.

37. Ulothrix tenissima Kutzing.

Plate-XXVII, Fig.37

Cells mostly 15-22μ thick (Seldom 14-30μ), in young stages as long as broad, but most often ½ - ½ the breadth, cell wall thin, not striated; chloroplast a broad band, with 2 or more pyrenoids.

Perceived free floating in water bodies of paddy field.

38. Ulothrix tenerrima (Kutzing) Kutzing.

Plate-XXVII, Fig.38

Cells 7-9 (-10)μ thick 2/3 - It times as long as broad; cell wall thin, most often somewhat mucilaginous; chloroplast gridle

shaped, encircling more than half the width of the cell, with one pyrenoid.
In soil cultures.

39. Ulothrix variabilis (Kutzing) Kutzing.

Plate-XWVII, Fig.39
Filaments sometimes with a pointed basal cell; cells 5-7μ broad, ½ - 1½ times as long as broad, cylindrical, with a thin cell wall; chloroplast usually not filling more than half the circumference of the cell and appearing as a plate covering one side of the cell with a small pyrenoid.
Perceived in algal collections as well as in soil cultures.

40. Ulothrix zonata. (Weber at Mohr) Kutzing

Plate-XXVII, Fig.40
Filaments attached in earlier stages, later free, with cylindrical or slightly swollen cells of varying width and length; length generally smaller than width; cells measure generally 11-45μ broad and 1/3 – 1½ times as long; cell wall thin in younger stages, but thickened in older filaments; chloroplast band shaped, broad or narrow gridle, covering only the median region of the cell and containing more than one pyrenoid, often several.
Perceived in algal collections from small water bodies of paddy field.

Genus - **Geminella** Turpienmend Lagerheim

41. Geminella minor (Naegeli) Heering

Plate-XXVIII, Fig.41
Cells cylindrical, 2-10μ broad, 2-4 times as long, always united together and surrounded by a mucilaginous sheath 8-18μ thick;

chloroplast parietal, covering part of the cell, with one or more pyrenoids.
In soil cultures.

42. Geminella protogenita. (Kutzing) G.S. West.

Plate-XXVIII, Fig.42
Cells cylindrical, 3.0-4.5μ broad, 1-2 times as long, filaments slightly constricted at the cross walls; cell wall thin; chloroplast longer, girdle shaped covering the wall and entire cell lengths; pyrenoid wanting.
In soil culture.

Family - Microsporaceae
Genus - **Microspora** Thuret.

43. Microspora abbreviate (Rabenhorst) Lagerheim.

Plate-XXVIII, Fig.43
Filaments 5-11μ broad; cells 1-3 times as long as broad; membrane homogeneous, yaline, somewhat thickened; chloroplast reticulate, perforate fully covering the wall; akinetes subspherical or elliptic.
Perceived free floating in water bodies of paddy field.

44. Microspora floccosa (Vaucher) Thuret.

Plate-XXVIII, Fig.44
Cells generally cylindrical, seldom constricted at the cross walls, 7.0-18μ broad, 2/3-2½ times as long; cell wall apparently thin, H-pieces usually not clear in vegetative condition; chloroplast pale green; often reticulate, or united in moniliform bands; akinetes 14.6-22μ in diameter.
Perceived in water bodies of paddy fields.

45. Microspora stagnorum (Kutzing) Lagerheim.

Plate-XXVIII, Fig.45

Filaments cylindrical, not constricted at cross walls; cells 5.0-10.0μ broad and 1-5 times as long; cell wall thin, the two halves generally indistinct, chloroplast variable even in the same filament, sometimes irregularly band like, more or less closely covering the wall around or forming a loose net work not completely filling the cell.

Perceived free floating in water bodies of paddy field.

Order	-	Cladophorales
Family	-	Cladophoraceae.
Genus	-	**Rhizoclonium** Kuetzing

46. Rhizoclonium heiroglyphicum. (Agardh) Kuetzing.

Plate - XXVIII, Fig. 46.

Filaments long, wiry, unbranched or very rarely branched, branches very short, hardly consisting of 2-4 cells; cells 28-35μ in diameter, 2-3 times longer than broad, (50-) 75-100μ long; chloroplast diffused, parietal reticulate, with many pyrenoids. Perceived in the algal collections from water bodies of paddy fields.

47. Rhizoclonium hookeri Kuetzing.

Plate - XXVIII, Fig. 47.

Filaments rarely branched, composed of long, cylindric, cells 55-70μ broad, 500-1000μ long; branch cells about the same diameter as those of the main filament or slightly narrower; secondary branching not observed.

Perceived in water bodies of paddy fields.

Order	-	Chaetophorales.
Family	-	Protococcaceae.
Genus	-	**Protococcus** Agardh.

48. Protococcus viridis Ag. (Prescott) Smith f.

Plate – XXVIII, Fig.48.

Solitary cell in spherical to ellipsoidal, bright green in colour with diffused chloroplast; diameter 8-20μ colonies of 2-4 cells are also found.

Perceived in soil cultures.

Family - Chaetophoraceae

Genus - **Chaetophora** Schrank

49. Chaetophora elegans (Roth) Ag.

Plate-XXVIII, Fig. 49

Colonies globose to flattened, soft, light blue-green in colour; main branches in the central part loose, rare and scarsely branched, branching dichotomus and fading loose fascicles in the peripheral region, ultimate branches short and without setae; cells 8-11μ in diameter, 15-22 μ long.

Perceived aligned to submerged paddy leaves and stalks.

50. Chaetophora pisiformis Ag.

Plate-XXVIII, Fig. 50

Colonies firm, globose to subglobose, light green in colour; main branches in the central portion of the mass loose, gradually ending in dichotomousely branched fascicles in the peripheral portion, ultimate branches long, setiferous; cells in the main filament 7-10μin diameter, 35-60μ long; branches narrower, cells of the branches (4-) 5-6μ in diameter.

Perceived aligned to submerged leaves and stalks of paddy.

Family - Coleochaetaceae

Genus - **Coleochaete** de Brebisson.

51. Coleochaete scutata de Brebisson.

Plate - XXIX, Fig. 51.

Thallus discoid with irregular outline; filaments compactly adjoined laterally, radiating from a common center; cells quadrangular to polygonal. 15-22μ in diameter, 12-30μ long.

Perceived aligned on submerged paddy leaves and stalks..

Order - Oedogoniales.

Family - Oedogoniaceae.

Genus - **Bulbochaete** Agardh.

52. Bulbochaete brevifulta Wittrock ex Hirn.

Plate - XXIX, Fig.52.

Vegetative cells cylindric, 15-22μ in diameter, 30-37μ long; oogonia broad ellipsoid, erect, 35-40μ in diameter, 50-55μ long; oospores with outer wall longitudinally and irregularly ribbed, 29-30μ in diameter, 45-48μ long.

Perceived in algal collections from water bodies of paddy field.

Genus - **Oedocladium** Stahl.

53. Oedocladium indicum Kamat.

Kamat, 1962 a, p.261, Pl.2, Fig.7.

Plate-XXIX, Fig.53

Monoecious; vegitative cells short cylindric to slightly torulose, 15-22μ in diameter, 20-30μ long, oogonia solitary, terminal, ellipsoid to ovoid, unripe oogonia, 27-32μ in diameter, 25-45μ long; antheridia terminal, 1, 10-12μ in diameter, 7.5-9μ long; antherozoids 2, division horizontal.

Perceived in algal collections from the water bodies of paddy fields.

Genus - **Oedogondium** link.

54. Oedogondium aquaticum Kamat.

Plate-XXIX, Fig.54

Dioecious, macrandrous; vegitative cells cylindric or very slightly capitellate, lone, mala filament cells 4-6μ in diameter, 25-30μ long, female filament cells slightly broader, 6-7μ in diameter, 22-34μ long; oogonia 2-4 seriate, broad ellipsoid, with rounded ends, 15-19μ in diameter 24-30μ long, poriferous, with a superior to nearly supreme pore; oospores of the same shape as oogonia, filling the oogonia, 14-18μ diameter, 22-28μ long, oospore wall smooth; antheridia 1-4 seriate, scattered or alternate with vegetative cells, 4.5-6μ in diameter, 6-9.5μ long, antherozoids two, division horizontal.

Perceived floating in water bodies of paddy field.

55. Oedogondium formosum Kamat

Plate - XXIX, Fig. 55.

Dioecious, macrandrous; vegitative cells, capitellate, 7.5-10μ in diameter, 30-48μ long; oogonia solitary, ellipsoid to obovoid, 20-22μ in diameter, 20-25μ long, operculate, operculum median; oospores of the same shape as oogonia, filling the oogonial 18-20μ in diameter, 21-23μ long; oospore wall smooth.

Perceived in algal collections from water bodies of paddy fields.

56. Oedogondium globosum Nordstedt ex Hirn.

Plate - XXIX, Fig. 56.

Monoecious; vegetative cells cylindric, 10.5-15μ in diameter, 30-38μ long; oogonium solitary, globose to sub globose, 32-40μ in diameter, 32-48μ long; poriferous, pore superior; oospores globose, filling the oogonia, rarely not filling the oogonial oospore wall smooth, antheridia 1-6. subepigynous or scattered, 11-13μ in diameter, 7 - 8μ long; antherozoids 2, division horizontal.

Perceived in water bodies of paddy field.

57. Oedogondium longipilum Jao.

Plate - XXIX, Fig.57.

Dioecious, macrandrous; filaments few celled, 8-10 celled, ending in a fine hair, epiphytic, vegitative cells not or slightly

capitellate; seta 600-800µ long; cells 6-6.5µ in diameter, 25-30µ long, oogonia solitary, intercalary, terminal, ellipsoid, 20-25µ in diameter 25-40µ long; operculate, with a supreme operculum; oospores ellipsoid ovoid, filling the oogonia, 19-20µ in diameter, 22-35µ long, oospore wall smooth.
Perceived aligned to submerged paddy leaves.

58. Oedogondium unisporum Kamat.

Plate-XXIX, Fig.58
Dioecious, macrandrous; vegitative cells cylindric, 3-4µ in diameter, 15-22µ long; oogonia solitary, ellipsoid, 10-12µ in diameter, 16-20µ long, operculate, with a median operculum, operculum wide; oospores broad ellipsoid with rounded ends to short cylindric, not filling the oogonia, 8-10µ in diameter, 11-13µ long, oospore wall smooth.
Perceived in water bodies of paddy fields.

59. Oedogondium maharastrensis Kamat

Plate-XXIX, Fig.59
Monoecious; vegitative cells cylindric, 15-18µ in diameter, 45-70µ long; oogonia 1-2 seriate, ellipsoid, obovoid to ellipsoid, 40-45µin diameter, 45-60µ long, operculate, operculum supreme; oospores globose, not filling the oogonia, 37-42µ in diameter, oospore wall smooth; antheridia scattered, 1-6 seriate, 10-14µ in diameter, 7-9µ long, antherozoids two, division horizontal.
Perceived in algal collections from water bodies of paddy field.

60. Oedogondium princeps (Hassall) Wittrock ex Him.

Plate - XXIX, Fig.60.
Dioecious; macradrous; vegitative cells cylindric, male vegetative cells 25-35µ in diameter, 60-120µ long, female vegitative cells 30-37µ in diameter, 60-100µ long; oogonia solitary, globose to ovoid, 60-75µ in diameter. 70-75µ long; poriferous, pore

superior) oospores globose, 55-60μ in diameter, 55-60μ long, not filling the oogonial oospore wall smooth; antheridia 15-20 in series, 22-25μ diameter, 4.5-6μ long.
Perceived in water bodies of paddy field.

61. Oedogondium rugulosum Nordstcdt ex Hirn.

Plate - XXIX, Fig. 6l.
Dioecious, nanandrous; filaments generally epiphytic 9-12 (-13) celled; end cell broadly rounded: basal cell elongate, club-shaped; 3-6 (-7)μ in diameter; upto 30μ long; oogonia solitary, ellipsoid, 19-22μ in diameter, 30-35μ long; operculate, with a superior to supreme operculum; oospores ellipsoid, obocoid; usually not filling the oogonia, rarely filling the oogonia; oospores 15-18μ in diameter; 25-30μ long; oospore wall smooth; nandria 2-4, straight, on the oogonia, 5-6.3μ in diameter, 10-15 μ long; antheridia solitary, 4-5μ in diameter, 5-6μ long.
Perceived in algal collections from water bodies of paddy fields.

Order	-	Zygnematales
Family	-	Zygnemaceae
Genus	-	**Spirogyra**. Link.

62. Spirogyra arta Jao.

Plate-XXX, Fig.62
Vegitative cell 15.5-18.6 to 80-140μ; zygospore 27-33 to 47-63 μ;
Perceived in algal collections from water bodies of paddy field.

63. Spirogyra biformis. Jao.

Plate-XXX, Fig.63
Vegitative cell 36-49 x 56-112 μ; zygospore 32-49 x 56-77 μ.
Perceived in algal collections from water bodies of paddy field.

64. Spirogyra communis (Hassal) Kutzing.

Plate-XXX, Fig.64

Vegitative cell 15-22 x 45-55μ; zygospore 19-25 x 36-60μ.

Perceived in algal collections from water bodies of paddy field.

65. Spirogyra chakiense Kolkwitz and Krieger.

Plate-XXX, Fig.65

Vegitative cell 91-105 x 98-112 μ; zygospore 48-63 x 8-126 μ.

Perceived in algal collections from water bodies of paddy field.

66. Spirogyra dictyospora Jao

Plate-XXX, Fig.66

Vegitative cell 28-31.5 x 63-140μ; zygospore 36-63 x 61-103μ.

Perceived in algal collections from water bodies of paddy field.

67. Spirogyra ghosei Singh

Plate-XXX, Fig.67

Vegitative cell 90-108 x 189-381 μ, zygospore 98-105 x 105-126 μ.

Perceived in algal collections from water bodies of paddy field.

68. Spirogyra hyaline Cleve

Plats-XXX, Fig.68

Vegitative cell 42 - 56 x 98 - 112 μ.

Zygospore 42-49 x 70-98 μ.

Perceived in algal collections from water bodies of paddy field.

69. Spirogyra irregularis Nageli.

Plate-XXX, Fig.69

Vegitative cell 35-42 x 154-190 μ

Zygospore 31-47 x 63-89.5 μ

Perceived in algal collections from water bodies of paddy field.

70. Spirogyra inflata (Vaucher) Kuetzing.

Plate-XXXI, Fig.70
Vegitative cell 17-19 x 80-200 μ
Zygospore 30-37 x 50-70 μ.
Perceived in algal collections from water bodies of paddy field.

71. Spirogyra maxima (Hassall) Wittrock

Plate-XXXI, Fig.71
Vegitative cell 125-154 x 175-292 μ.
Zygospore 70-98 x 110-140 μ.
Perceived in algal collections from water bodies of paddy field.

72. Spirogyra prolifica Kamat.

Plate-XXXI, Fig.72
Vegitative cell 25-32 x 120-200 μ.
Zygospore 40-50 x 78-100 μ.
Perceived in algal collections from water bodies of paddy field.

73. Spirogyra submaxima Transeau.

Plate-XXXI, Fig.73
Vegitative cell 90-120 (-140) x (60-) 125-220 μ.
Zygospore (65-) 75-100 x 50-75 μ.
Perceived in algal collections from water bodies of paddy field.

74. Spirogyra turfosa Gay.

Plate-XXXI, Fig.74
Vegitative cell 70-84 x 126-294 μ.
Zygospore 63-77 x 70-200 μ.
Perceived in algal collections from water bodies of paddy field.

Genus - **Sirocladium** Randhawa

75. Sirocladium maharashtranse Randhawa

Plate-XXXI, Fig.75

Vegitative cells cylindrical with plane end walls; 37.5-41.25μ broad, 108.75-207.5μ long; chloroplasts 2; parietal ribbons 9.0-11.2μ broad, each with 4-14 pyrenoids arranged in a row. Perceived in algal collections from water bodies of paddy fields.

Family - Mougeotiaceae.
Genus - **Sirogonium** Kuetzing.

76. Sirogonium ventersicum Transeau.

Plate - XXXI, Fig.76.

Vegitative cells 63-75μ broad, 190-320μ long, with plane end walls; chloroplasts 4, straight or nearly straight, conjugation direct, receptive gametangia slightly inflated, 100-120μ broad; zygospores ovoid, 80-95μ broad, 120-180μ long, mesospore irregularly verrucose.

Perceived in water bodies of paddy field.

Family - Desmidiaceae.
Genus - **Closterium** Nitzsch.

77. Closterium acerosum (Schrank) Ehrenberg.

Plate - XXXII, Fig.77.

Cells large, fairly or slightly curved or almost straight, narrowly fusiform, outer margin slightly curved. gradually attenuated with rounded apices, 21-55μ in diameters, 250-500μ long, cell wall smooth colourless, with one or more median girdles; chlorcplast ridged; pyrenoids 5-8 in a row.

Perceived in the algal collections from water bodies of paddy field as well as in soil cultures.

78. Closterium acutum. (Lyngbye) de Brebisson.

Plate - XXXII, Fig.78.

Cells slightly curved or straight, gradually tapering towards acute apices; inner margin not tumid; cell wall colourless, 7.0-

12.5μ in diameter; 75-170μ long, chloroplast parietal without longitudinal ridges and with 3-4 pyrenoids.

Perceived in algal collections from water bodies of paddy fields.

79. Closterium lanceolatum Kuetzing.

Plate - XXXII, Fig.79.

Cells long, sub-lanceolata, curved, 7-10 times longer than broad, 40-50 in diameter, at the apices 7-9μ broad, 300-425μ long, gradually tapering at the ends, ends acute, rounded; outer margin slightly curved, 30-55 degrees of arc, inner margin straight, or slightly convex; cell wall smooth, colourless; chloroplast with 7-8 ridges, with 7-10 pyrenoids in a median series.

Perceived in algal collections from water bodies of paddy fields.

80. Closterium paryulum Nag.

Plate - XXXII, Fig.80.

Cells fairly curved, attenuated towards the apices; 10-15μbroad and 125-200μ long, cells wall smooth colourless; chloroplast plate like with 3-4 pyrenoids each.

Perceived in algal collections from much wet soil surface of paddy field as well as in soil cultures.

81. Closterium venus Kuetzing.

Plate - XXXII, Fig.8l.

Cells fairly curved, attenuated towards acute apices 7-12μ broad 100-125μ long; cell wall smooth, colourless, chlorcplast platc likc with 2 pyrcnoids cach.

Perceived in algal collections from wet soil surfaces of paddy fields.

Genus - **Cosmarium** Corda

82. Cosmarium contractum Kirchner.

Plate - XXXII, Fig.82.

Cells slightly longer than broad, 20-25μ broad, 30-34μ long; isthmus 6-8μ broad, deeply constricted, sinus narrow towards the apex, widening outwards; semicells sub- elliptic; circular inside view; elliptic in vertical chloroplast axile, with a pyrenoid.

Perceived in water bodies of paddy field.

83. Cosmarium depressum (Naeg.) Lund. Var. **Achondrum** (Boldt) West et west.

Plate XXXII, Fig.83.

Cells a little broader than long, 50-53μ broad, 40-45μ long, isthmus 18-22μ broad, deeply constricted, sinus narrowly linear, opening outwards; semicells subhexagonal elliptic, apex broader and truncate; side view of semicells circular, vertical view elliptic; cell wall smooth; chloroplast axile, with a pyrenoid.

Perceived in water bodies of paddy field.

84. Cosmarium dentiforum Corda Var. **minus** Kamat.

Plate - XXXIII, Fig. 84.

Cells as long as broad or very slightly longer than broad, 50-54μ in diameter, 52-60μ long; isthmus 13-16μ broad, deeply constricted, sinus narrow and dilated at extremity; semicells reniform with a slightly depressed apex; vertical view oblong, poles rounded; cell wall granulate, granules rounded; chlorcplast 2, axile with 2 pyrenoids in each semicells.

Perceived in wet soil surfaces of paddy fields.

85. Cosmarium granatum Brebisson.

Plate - XXXIII, Fig.85.

Cells longer than broad, upto 1½ times as long as broad, 18-25μ in diameter, 22.40μ long; isthmus 5-10μ in diameter, subrhomboidal eeliptic, deeply constricted, sinus narrow, linear, slightly dilated at apex; semicells truncate and straight, side view of semicells elliptic cell wall smooth; chloroplast axils, with a central pyrenoid.
Perceived in algal collection from wet soil surfaces as well as in soil cultures.

86. Cosmarium impressulum Elfv.

Plate - XXXIII, Fig.86.
Cells longer than broad, small, 18-22μ in diameter, 22-25μ long, isthmus 4-5μ in diameter, 10-15μ thick; deeply constricted, sinus narrowly linear, with a slightly dilated apex; semicells elliptic to subsemicircular, margin with 8 undulations; vertical view elliptic; cell wall smooth, chloroplast axile, with a central pyrenoid.
Perceived on wet soil surfaces of paddy fields.

87. Cosmarium meneghinii de Brebisson.

Plate - XXXIII, Fig. 87.
Cells longer than broad, about 1½ times as long as br0ad, deeply constricted, sinus narrow and linear, dilated at the apex; semicells transversely rectangular in the lower part, and pyramidate in the upper part, apex retuse, side view of semicells broadly elliptic, vertical view elliptic; cells 23-25μ broad; 28-33μ long, isthmus 7-9μ broad; cell wall smooth, chlorcplast axile. with a pyrenoid.
Perceived in water bodies of paddy field.

88. Cosmarium subimpressulum. Borge.

Plate XXXIII, Fig.88.

Cells small, longer than broad, deeply constricted, 16-20μ in diameter, 25-30μ long, isthmus 6.3-7μ in diameter, sinus linear, narrow; semicells transversely truncate, crenate, with 4 crenations on each side, in part with broad inflations on each side, lateral view of semicells ovate, apex truncate, tumid in the middle portion on both the sides; cell wall smooth; chloroplasts solitary, with a central pyrenoid.

Perceived in water bodies of paddy field.

89. Cosmarium trilobatum Reinsch.

Plate - XXXIII, Fig.89.

Cells small, longer than broad upto 1½ times as long as broad, deeply constricted, sinus narrowly linear, without or with slightly dialated extremity, semicells somewhat three lobed, lobes short, subrectangular with rounded angles; incisions between lobes wide and shallow; side view of semicells broadly semi elliptic., vertical view elliptic, 12-13μ broad, 18-20μ long, esthmus, 3-4μ broad; cell wall smooth, chloroplast axile with a pyrenoid.

Perceived in water bodies of paddy field.

90. Cosmarium undulatum Corda ex Ralfs.

Plate-XXXIII, Fig. 90

Cells large, slightly longer than broad, deeply constricted, 45-52μ in diameter, 52-60μ long; isthmus 12.16μ in diameter, sinus linear and open, dialated at apex; semicells subcircular in side view, vertical view elliptic; cell wall smooth with undulations; chloroplasts 2, each with a pyrenold.

Perceived on wet soil surfaces of paddy field.

Division	-	Xanthophyta
Class	-	Xanthophyceae
Order	-	Siphonales

Family - Vaucheriaceae
Genus - **Vaucheria**

1. Vaucheria amphibian Randhawa

Plate-XXXIV, Fig. 1
Filaments 42-66μ broad.
Oogonia in pairs on short laterial branches, surrounding a terminal antheridium, occasionally solitary and pushed to one side with an antheridium on the opposite side, globose to pear-shaped, 72-110μ broad, 80-120μ long; oospores completely filling the oogonium, 70-106μ broad, 78-116μ long, dark in colour, antheridia terminal or lateral, slightly curved, 30-44μ broad.
Perceived forming green patch on soil surfaces.

2. Vaucheria geminata (Vauch.) De Condolle

Plate-XXXIV, Fig.2
Filaments 35-110μ broad.
Oogonia one to six, 60-110μ, broad, 70-120μ long, elipsoid hemispherical to convex-concave, shortly pedicelled near the end of a short branch; oospore filling the oogonium, 52.5-225μ broad, 64.5-190μ long, membrane three-layered; antheridia cylindrical, hooked or circinate, usually on a stipe longer than those of the oogonia; aplanosporangia on the same filament as the oogonia or on separate filaments; aplanospores 120-200 μ broad, 120-190 μ long, aplanosporangia ovoid, terminating short lateral branches, dissemination by the dissolution of the membrane; akinetes formed by breaking up of portions of the filaments into short thick walled cells.
Perceived in algal mat collections from wet soil surfaces.

3. Vaucheria hamata Walz

Plate-XXXIV, Fig. 3

Filaments 38.5 to 55 μ broad.

Oogonia one to two, ovoid to convex-concave, borne on the shorter division of an apparently forking branch, the longer division recurved bearing the hooked or circinate antheridia or an oogonium on each division; oospore filling the oogonium, 55-66 μ broad, 66-88 μ long, membrane four layered; grayish in colour; antheridium in a place at an angle to that of the oogonium, 12-21 μ broad; aplanosporangia terminal, clavate.

Perceived on wet soil surfaces.

4. Vaucheria mayyanadensis Erady

Plate-XXXIV, Fig. 4

Filaments 25-97 μ broad, sparsely branched; rhizoidal filaments colourless, 7-62 μ broad.

Dioecius; oogonia one to two or rarely more, sessile, seldom shortly staked, 103-178 μ broad, 133-223 μ long spherical or ovated; long axis making an angle with or seldom parallel to or at right angle to the erect filament; break hyaline; oospores spherical, not completely filling the oogonium; oospore membrane thick, spotted, 89-149 μ broad; antheridia solitary or in clusters, sessils, ovate or broadly lanceolate, lateral or terminal at the end of the short or long branches, 44-75 μ broad, 103-193 μ long; terminal break blunt.

Perceived on wet soil surfaces.

5. Vaucheria sessilis (Vauch) De Condolle

Plate_XXXIV, Fig. 5

Filaments 55-110 μ broad.

Oogonia usually two, sometimes one, mostly sessile, seldom on short pedicles, ovate or oblong-ovate, 57-77 μ broad, 80-102 μ long, slightly oblique with a short oblique beak; antheridium behind a single oogonium or between 2 oogonia on a short straight or slightly curved pedicel, hooked to circinate, 28-32 μ broad; oospore filling the oogonium, dark-spotted, membrane three layered, 51-71 μ broad, 73-96 μ long; zoosporangia ovoid-clavate, terminal on a branch, zoospores 83-176 μ long, 77-157 μ broad, cilia uniformly distributed.

Perceived on wet soil surfaces.

Order	-	Heterosiphonales
Family	-	Botrydiaceae
Genus	-	**Botrydium**

6. Botrydium species

Plate-XXXIV, Fig. 6

Bladder like body has upper pigmented blund portion and lower colourless narrowed portion. Upper portion measures 84-86μ in diameters and narrows towards the lower portion. The lower portion is 25-30μ in diameter. The cytoplasm contains several scattered nuclei and several discoid chromatophores which are often connected with one another by strains of dense cytoplasm.

In soil cultures.

Division	-	Euglenophyta.
Class	-	Euglenophyceae.
Order	-	Euglenales.
Family	-	Euglenaceae.
Genus	-	**Euglena**

1. Euglena acus Ehrenberg v. **hyalina** Klebs.

Plate - XXXV, Fig. l.

Cells elongate, spindle shaped, produced posteriorly into a long, fine tapering point, narrowed and truncate at the anterior end; periplast indistinctly spirally striated; chromatophores absent,

paramylon bodies 4-10 in the form rods; cells 10-11μ broad, 145-165μ long.
Perceived in algal collections from water bodies of paddy field.

2. Euglena charkowiensis Swir

Plate - XXXV, Fig.2.
Cells long, often twisted, rounded at the posterior end with a pointed and usually curved tail rounded at the anterior end; not metabolic; periplast distinctly striated; chromatophores many, disc-shaped, pyrenoids not observed, paramylon bodies 2, cylindrical; cells 15-30 (-50)μ broad, 75-300μ long.
Perceived in algal collections from water bodies of paddy field.

3. Euglena gracillis Klebs

Plate - XXXV, Fig.3.
Cells fusiform to avoid, with or without a pointed posterior end; metabolic; chromatophores many, discoid, evenly distributed throughout the: cell, with pyrenoid bodies, paramylon bodies not observed; cells 16-30μ broad, 60-75μ long.
Perceived intermingled with other free floating algae in water bodies of paddy field.

4. Euglena multiformis Schiller

Plate – XXXV, Fig. 4.
Cells fusiform, tapering, in the posterior part; metabolic; periplast distinctly striated, flagellum shorter than the body length; chromatophores many, disc-shaped, paramylon bodies globose, ring-like; cells 6-10μ broad, 30-50μ long.
Perceived in water bodies of paddy field.

5. Euglena tuberculata swirenko.

Plate - XXXV, Fig. 5.

Cells broadly spindle shaped, tapering into a short, straight tail; periplast finely striated, striations indistinct, tuberculate; chromatophores many, small, discoid, paramylong bodies not observed; cells 12-15μ broad, 40-45μ long.

Perceived in water bodies of paddy field.

6. Euglena fusca (Klebs) Lemm.

Plate - XXXV, Fig. 6.

Cells ellipsoid, tapering gradually in a tail; mostly keeping a constant shape in the movement; periplast studied with rectangular, spirally arranged markings; chramatophores many, disc-shaped, paramylon bodies 2, ring-like, cells 22-30μ broad, 125-200μ long.

Perceived in water bodies of paddy field.

7. **Euglena maharastrensis** Kamat.

Plate - XXXV, Fig.7.

Cells ellipsoid, slightly bent, slightly tapering in the posterior part, 50-70μ broad, 5-6μ long, chromatophores many, discoid.

Perceived in water bodies of paddy field.

Genus - **Phacus.**

8. Phacus acuminatus stokes v. **discifera** (Pochm) Huber - Pestalozzi.

Plate - XXXV, Fig.8.

Cells broadly ovoid, broader in the posterior part: and slightly narrower in the anterior part; posterior part suddenly tapering into a wedge shape tail; periplast longitudinally striated; chromatophores many, small, discoid, paramylon bodies 2, discoid, one small and the other big; cells 28-30μ broad, 35-37μ long.

Perceived in algal collections from water bodies of paddy field.

9. Phacus caudatus Huebner.

Plate - XXXV, Fig. 9.

Cells ovoid, flattened, often slightly twisted, posterior part suddenly tapering to form a short, straight tall; broadly rounded anterior end with a distinct notch in the anterior end; periplast longitudinally striated; chromatophores many, small, discoid; paramylon bodies large, disc-like; cells 28-31μ broad, 46-50μ long.

Perceived in water bodies of paddy field.

10. Phacus hamatus Pochmann.

Plate - XXXV, Fig. 10.

Cells long, ellipsoid, narrowing at both the ends, slightly broader in the posterior part suddenly tapering into a short tail; periplast longitudinally striated; chromatophares many, disc shaped, paramylon body a disc in the posterior part; cells 25-30μ broad, 40-42μ long. , Perceived in water bodies of paddy field.

11. Phacus longicauda (E) Dujardin.

Plate - XXXV, Fig.11.

Cells ovoid to pyriform, posterior end tapering in a long, straight, sha:rply pointed tail, anterior broadly rounded; periplast longitudinally striated; flagellum shorter than the body length; chromatophores many, discoid, paramylon body 1, large disc; cells 60-70μ broad, 90-100μ long.

Perceived in water bodies of paddy field.

12. Phacus platalea Drez. v. **minor** Kamath.

Plate - XXXV, Fig.12.

Cells broadly oval, flattened, with posterior part tapering suddenly to a pointed, slightly bent tail; periplast longitudinally

striated; chromatophores many, small. Disc-shaped, paramylon body solitary, circular, ring-like, cells 20-30μ broad, 35-38μ long.
Perceived in water bodies.

13. Phacus horridus Pochmann

Plate - XXXV, Fig. 13.
Cells rectangular spoon-shaped, oval, ending in a sort tail, cell wall beset with bent, pointed spines; flagellum longer than body length; periplast longitudinally striated; chromatophores many, discoid, cells 25-28μ broad, 40-45μ long.
Perceived in water bodies.

14. Phacus curvicauda Swirenko.

Plate XXXV, Fig. 14.
Cells broadly oval to suborbicular, slightly spiral in the posterior part, posterior part ending in a tail; anterior end broadly rounded; per.iplast longitudinally striated; chromatophores many, oval, paramylon bodies 2, disc-like; cells 25-32μ broad, 30-35μ long.
Perceived in water bodies.

15. Phacus tortus (Lemm.) Skvortzov.

Plate - XXXV, Fig.15.
Cells broad ovoid, napiform, broadest in the middle portion, conically rounded in the anterior portion, suddenly tapering in the posterior part, spirally twisted in the posterior portion, ending into a long, pointed tail: periplast striated; chromatophores many, paramylon body 1, ring-like disc; cells 25-35μ broad, 80-90μ long.
Perceived in water bodies.

Genus - **Trachelomonas.**

16. Trachelomonas allia Drez. em. Deflandre.

Plate - XXXVI, Fig.16.

Test cyclindric - ellipsoid, with broad rounded poles, with two sides parallel for half the length, dark brown; test beset with fine, short spines; flagellum aperture with a very short collar beset with spines; test 20-22µ broad, 28-30µ long.

17. Trachelomonas hispida (Perty) Stein emnd. Deflandre.

Prescott, 1951, p. 414, p.

Plate - XXXVI, Fig. 17.

Test ovoid. narrowed anteriorly. dark brown to red brown; flagellum aperture slightly raised; uniformly beset with mmute shaz:p I pomted spmes; test 12-16µ in diameter, 18-20µ long. Perceived in algal collection s fran water bodies.

18. Trachelomonas lefevrei Deflandre.

Test ellipsoid to ovoid, broader in the anterior part, yellow brown to dark brown in colour, beset with thick punctae; flagellum aperture with a short, cylmdric collar; test 18-20µ broad, 25-30µ long.

Perceived in algal collections from water bodies.

19. Trachelomonas oblonga Lemmermann.

Plate - XXXVII Fig.459.

Test ellipsoid. snoath, hyaline to yellow, flagellum aperture with or without collar, chromatophores 2; in diameter, 12-14µ long.

Perceived in algal collections from water bodies.

20. Trachelomonas volvocina Ehrenoerg.

Plate XXXVI, Fig.20.

Test globose, snooth, yellow-brown to dark brown; flagellum aperture with or without a collar; test 18-21µ in diameter.

Perceived in algal collections from water bodies.

21. Trachelomonas volvocina Ehrenberg v. **derephora** Conrad.

Plate - XXXVI, Fig.21.
Test spherical, smooth, hyaline to red brown; flagellum aperture with a cylindric well developed collar; test 12-13µ broad.
Perceived in algal collections from water bodies.

22. Trachelomonas armata (E) stein v. **steinii** emend. Deflandre.

Plate - XXXVI, Fig. 22.
Test ellipsoid, ovoid slightly broader in the posterior part, brown, lines allover; spines of two sizes, smaller all over the body, bigger spinesrestricted to the posterior part, collar beset with smaller spines; collar ring present; tests 30-35µ in diameter, 40-50µ long.
Pervceived in algal collections from water bodies.

23. Trachelomonas cordata (Drez.) comb. Deflandre f. **minor** Deflandre.

Plate - XXXVI, Fig.23.
Test ellipsoid to ovoid, red brown, with a short collar, 8-10µ in diameter, 10-12µ long.
Perceived in algal collections from water bodies.

24. Trachelomonas hispida (Perty) Stein emend. Deflandre f. **minor** Bourr.

Plate - XXXVI, Fig.24.
Test ovoid, brown to red brown, narrowed anteriorly, wall uniformly beset with sharp, pointed spines, with narrow collar; test 15-16µ in diameter, 19-22µ long, spines 1-1.5µ long.
Perceived in algal collections from water bodies.

25. Trachelomonas pulcherrima Playfair.

Plate - XXXVI, Fig. 25.

Test ellipsoid, red to red brown, smooth, with a collar; flagellum longer than the body length; test 9-16μ in diameter, 18-25μ long.

Perceived in algal collection from water bodies of paddy field.

Division	-	Bacillariophyta
Class	-	Bacillriophyeeae
Order	-	Centrales
Family	-	Coscinodiscaeeae
Sub-family	-	Coscinodicoideae
Genus	-	**Cycllotella** kutzing

1. Cyclotella kutzingiana Thwaites

Venkataraman, 1939, P. 299, F. 9

Plale- XXXVII , Fig. 1

Dimensions: Diameter 25.3 - 30 μ
Striae12-13 in 10μ

Perceived in algal collections from water bodies of paddy field.

2. Cyclotella operculata Kg.

Plate-XXXVII, Fig.2

Dimensions: Diameter 11.2 - 13.2
Striae 13 - 14 in 10 μ

Perceived in algal collections from water bodies of paddy field.

Order	-	Pennales
Sub-order	-	Araphidineae
Family	-	Fragillariaeeae
Sub-Family	-	Fragillarioideae
Genus	-	**Fragillaria** Lyngbye

3. Fragillaria brevistriata f. elongata Venk.

Plate-XXXVII, Fig.3

Dimensions : Length 32-35μ
Breadth 4.2-5.6 μ
Striae 12-16 in 10 μ

Perceived in water bodies of paddy field as well as in cultures.

Genus - **Synedra** Ehrenb.

4. Synedra affinis. Kg.

Plate-XXXVII, Fig.4

Dimensions: Length 80-115.0μ
Breadth 4.2 - 5.6μ
Straie 12-13 in 10μ

Perceived in water bodies of paddy field as well as in cultures.

5. Synedra pulchella (Ralfs.) Kg. var. **lanceolata** O. Meara

Plate-XXXVII, Fig.5

Dimensions: Length - 50-58.8μ
Breadth - 7.0-8.4μ
striae - 14-16 in 10μ

Perceived in water bodies of paddy field as well as in cultures.

6. Synedra ulna (Nitzsch) Ehrenb.

Plate-XXXVII, Fig.6

Dimensions : Length 91-82μ
Breadth 4.2-4.8μ
Striae 10-16 in 10μ

Perceived in water bodies of paddy field as well as in cultures.

Sub order - Monoraphidineae
Family - Achnathaeeae
Sub-family - Achnathoideae
Genus - **Achnanthes** Bory.

7. Achnanthes coarctata Breb.

Plate-XXXVII, Fig.7

Dimensions: Length 37.8-42μ
Breadth 7.0-8.4μ
Straie 11-13 in 10μ

Perceived in water bodies of paddy field as well as in cultures.

8. Achnanthes delicatula Kg.

Plate-XXXVII, Fig. 8

Dimensions: Length 16.8-20μ
Breadth 8.4-9.8μ
Striae 14-15 in 10 u
Striae 14-15 in 10μ

Perceived in water bodies of paddy field as well as in cultures.

9. Achnanthes microcephala Kg.

Plate-XXXVII, Fig. 9

Dimensions: Length 22.4-22.8μ
Breadth 1.4-2.8μ
Striae 30-32 in 10μ

Perceived in water bodies of paddy field as well as in cultures.

Genus - **Gyrosigma** Hassall

10. Gyrosigma spenceri (W. Sm.) Cl.

Plate-XXXVII, Fig. 10

Dimensions: Length 79.5-108μ
Breadth 12.7-16.2μ
Striae 17-20 in 10μ

Perceived in water bodies of paddy field as well as in cultures.

Genus - **Stauroneis** Cleve

11. Stauroneis anceps Ehrehb.

Plate-XXXVII, Fig. ll

Lund, 1946, P.59, Fig.3 I, J.

Dimensions: Length 35-43.2 µ
Breadth 7.8-8.4µ
Striae 19-20 in 10µ

Perceived in water bodies of paddy field as well as in cultures.

Genus - **Navicula** Bory

12. Navicular anceps (Ehrenb.) var. **linearis** Ehrenb.(als Art)

Plate- XXXVII , Fig. 12

Dimensions: Length 30-44.2 µ
Breadth 7.0-13.2 µ
Striae 20-25 in 10 µ

Perceived in water bodies of paddy field as well as in cultures.

13. Navicula aponina Kg.

Plate-XXXVII, Fig.13

Dimensions: Length 25.3-26.7 µ
Breadth 4.2-5.6 µ
Striae 30 in 10 µ

Perceived in water bodies of paddy field as well as in cultures.

14. Navicula bacilliformis Grun.

Plate-XXXVII, Fig.14

Dimensions: Length 37.8-48.0 µ
Breadth 9.8-12.6µ
Striae in the middle 12-16 in 10µ
towards end 20-22 in 10µ

Perceived in water bodies of paddy field as well as in culture.

15. Navicula brebissonii Kg.

Plate-XXXVII, Fig.15

Dimensions: Length 57.4-66.0 µ
Breadth 9.8-14.2 µ
Striae 10-12 in 10 µ

Perceived in water bodies of paddy field as well as in cultures.

16. Navicula clavata Greg.

Plate-XXXVII, Fig.16

Dimensions: Length 64.5-76.0μ
Breadth 22.4-25.2μ
Striae 10-14 in 10μ

Punctae 16-20 in cultures.

17. Navicula crucicula (W. Sm.) Donk.

Plate-XXXVII, Fig.17

Dimensions : Length - 39-54μ
Breadth - 11.4-22.2μ
Striae - 16-18 in 10μ

Perceived in water bodies of paddy field as well as in cultures.

18. Navicula cryptocephala Kg. .f. **terrestris** Lund

Plate-XXXVII, Fig.18

Dimensions: Length - 18.9-24.0μ
Breadth - 4.2-6.0μ
Striae - 15-20 in 10μ

Perceived in water bodies of paddy field as well as in cultures.

19. Navicula gregaria Donk.

Plate-XXXVII, Fig.19

Dimensions: Length - 23.8-28.2μ
Breadth - 7.0-9.6μ
Striae - 17-18 in 10μ

Perceived in water bodies of paddy field as well as in cultures.

20. Navicula grivillei (Ag.) Cl.

Plate – XXXVII, fig. 20

Dimensions : Length - 50-60μ
Breadth - 15-17μ
Striae - 8-10 in 10μ

Perceived in water bodies of paddy field as well as in cultures.

21. Navicula lanceolata (Ag.) Kg.

Plate-XXXVIII, Fig.21

Dimensions: Length - 37.8-42.6μ
Breadth - 8.4-9.8μ
Striae - in the centre 12 in 10μ
toward the end 15-21 in 10μ

Perceived in water bodies of paddy field as well as in culture.

22. Navicula major Kg.

Plate-XXXVIII, Fig.22

Dimensions: Length - 200-216μ
Breadth - 200-300μ
Striae - 3-5.5 in 10μ

Perceived in water bodies of paddy field as well as in cultures.

23. Navicula mutica Kutz.

Plate-XXXVIII, Fig. 23

Dimensions Length - 15-36 μ
Breadth - 5.6-7.8 μ
Striae - 18-20 in 10 μ

Perceived in water bodies of paddy field as well as in cultures.

24. Navicula sculpta Ehrenb.

Plate-XXXVIII, Fig.24

Dimensions: Length - 98.3-108 μ
Breadth - 25.3-28.2μ
Striae - 15-16 in 10 μ

Perceived in water bodies of paddy field as well as in cultures.

25. Navicula salinarum Grun.

Plate-XXXVIII, Fig.25

Dimensions: Length - 35-41.6μ
Breadth - 9-11μ
Striae -12 in 10μ

Perceived in water bodies of paddy field as well as in cultures.

26. Navicula terrestris Petersen.

Plate-XXXVIII, Fig.26

Dimensions: Length – 29.4-40.8µ
Breadth – 2.8-13µ
Striae -13 in 14 in 10µ
toward end 18-20 in 10µ

Perceived in water bodies of paddy field as well as in cultures.

27. Navicula tennelloides (Hust.) emend.

Plate-XXXVIII, Fig. 27.

Dimensions : Length – 14.4 µ
Breadth – 4.0-4.8 µ
Striae – not clearly visible.

Perceived in water bodies of paddy field as well as in cultures.

28. Navicula ucheriae var. **densistriata** Lund.

Plate - XXXVIII; Fig.28.

Dimensions: Length – 14-16.8µ
Breadth – 4.2-6.0µ
Striae – in the middle 18-20 in 10µ
towards the ends 30-32 in 10 µ

Perceived in water bodies of paddy field as well as in culture.

Genus - **Pinnularia** Threnberg.

29. Pinnularia microstauron (Ehr.) Cleve.

Plate - XXXVIII, Fig.29.

Dimensions: Length – 30.8-36.0µ
Breadth – 5.6-9.8µ
Striae -12-15 in 10µ

Perceived in water bodies of paddy field as well as in cultures.

30. Pinnularia viridis (Nitzsch.) Ehr.

Plate-XXXVIII, Fig.30

Dimensions : Length - 56.1-63.1 μ
Breadth - 15.4-16.8μ more breadth than the type.
Striae - 7-8 in 10μ

Perceived in water bodies of paddy field as well as in cultures.

Genus - **Amphora** Cl.

31. Amphora caffaeiformis Ag.

Plate-XXXVIII, Fig. 31

Dimensions : Length - 31.8-35.8 μ
Breadth - 12.7-16.2 μ
Striae - 20-21 in 10 μ

Perceived in water bodies of paddy field as well as in cultures.

32. Amphora ovalis Kg.

Plate-XXXVIII, Fig.32

Dimensions : Length - 36.4-44.4 μ
Breadth - 15.4-21.0 μ
Striae - 11-13 in 10 μ

Perceived in water bodies of paddy field as well as in cultures.

33. Amphora Montana (Krasake)

Plate-XXXVIII, Fig.33

Dimensions : Length - 18-20 μ
Breadth - 5-6 μ
Striae - 16 in 10 μ

Perceived in water bodies of paddy field as well as in cultures.

Genus - **Cymbella** Ag.

34. Cymbella alpine Grun.

Plate-XXXVIII, Fig.34

Dimensions: Length - 39.8-43.8 μ
Breadth - 8.4-9.8 μ
Striae - 5-7 in 10 μ

Perceived in water bodies of paddy field as well as in cultures.

35. Cymbella austriaca Grun.

Plate-XXXIX, Fig.35

Dimensions: Length - 43.8-49 μ
Breadth – 11–13 in 10 μ

Perceived in water bodies of paddy field as well as in cultures.

36. Cymbella cymbiformis (Kg.) V.H.

Migula, 1907-09, P.299, Taf. XE, Fig.4

Plate-XXXIX, Fig.36

Dimensions: Length - 63-73μ
Breadth - 11.2-13.2μ
Striae - 8-9 in 10μ
Punctae - 20 in 10 μ

Perceived in water bodies of paddy field as well as in cultures.

37. Cymbella gracilis (Rabenh.) Cleve

Migula, 1907-09, P.305, Taf. XE, Fig.9

Plate-XXXIX, Fig.37

Dimensions: Length - 39.2-44.4 μ
Breadth - 8.4-11.4 μ
Striae 10-15 in 10 μ

Perceived in water bodies of paddy field as well as in cultures.

38. Cymbella parva W. Sm.

Plate-XXXIX, Fig.38

Dimensions: Length - 37.8-40.5μ
Breadth - 9.8-11.2μ
Striae - 9-12 in 10μ

Perceived in water bodies of paddy field as well as in cultures.

39. Cymbella reinhardtii Grun.

Plate-XXXIX, Fig.39

Dimensions : Length – 41.2-45.4 μ
Breadth – 9.5-14.8 μ
Striae – in the middle 10-11 in 10μ
towards th ends of 14 in 10 μ

Perceived in water bodies of paddy field as well as in culture.

40. Cymbella tumescens. A. Cl.

Plate-XXXIX, Fig.40

Dimensions: Length - 28.8-32μ
Breadth - 8-9.4μ
Striae - 11-12 in 10 μ

Perceived in water bodies of paddy field as well as in cultures.

41. Cymbella ventricosa (Kg.) Cleve.

Plate-XXXIX, Fig.41

Dimensions: Length 30.8-39μ
Breadth - 9.8-12μ
Striae - 9-12 in 10μ

Perceived in water bodies of paddy field as well as in cultures.

Family - Epithemiaceae
Sub-family - Rhopalodiodeae
Genus - **Rhopalodia** O. Mull

42. Rhopalodia gibba (Ehrenb.) O. Muller Var. **Ventricosa** (Ehrenb.) Grun.

Plate-XXXIX, Fig.42

Dimensions: Length - 44.8-50.4μ
Breadth - 9.8-11.2μ
Striae - 5-6 in 10 μ

Perceived in water bodies of paddy field as well as in cultures.

Genus - **Gomphonema**, Ag.

43. Gomphonema angustatum (Kg) Grun. Var. **producta** Grun.

Plate-XXXIX, Fig.43

Dimensions : Length - 20-24μ
Breadth - 7.0-7.4μ
Striae - 10-12 in 10μ

Perceived in water bodies of paddy field as well as in cultures.

44. Gomphonema gracile Ehrenb.

Plate-XXXIX, Fig.49

Dimensions : Length - 56-64 μ
Breadth - 9.8-12 μ
Striae - 12-13 in 10 μ

Perceived in water bodies of paddy field as well as in cultures.

45. Gomphonema lanceolatum Ehr.

Venkataraman, 1939, P.347, Fig.87

Plate-XXXIX, Fig.45

Dimensions : Length - 44-45 μ
Breadth - 12-13 μ
Striae - 10-11 in 10 μ

Perceived in water bodies of paddy field as well as in cultures.

Family - Nitzschiaeeae
Sub-family - Nitzschcideae
Genus - **Hantzschia** Grun.

46. Hantzschia anphioxys Grun

Lund, 1946, P.96, Fig.13 A-D. F.

Plate-XXXIX, Fig. 46

Dimensions : Length - 45-57.6 μ
Breadth - 8.4-9.8 μ
Striae - 21-22 in 10μ

Carinal dots - 3-4 in 10μ

Genus - **Nitzschia** Hass

47. Nitzschia angustata (W. Sm.) Grun.

Plate-XXXIX, Fig.47

Dimensions: Length - 71.5-90 μ
Breadth - 10.8-13.5 μ
Striae - 9-10 in 10 μ

Perceived in water bodies of paddy field as well as in cultures.

48. Nitzschia dissipata (Kg.) Grun.

Plate-XXXIX, Fig.48

Dimensions: Length - 28.4-33 μ
Breadth - 4.2-5.6 μ
Striae - 14 in 10 μ
Keel punctae 7-9 in 10 μ

Perceived in water bodies of paddy field as well as in cultures.

49. Nitzschia gracilis Hantzsch

Plate-XXXIX, Fig.49

Dimensions : Length - 84.2-92.4 μ
Breadth - 4.2-5.6 μ
Striae - 25-26 in 10 μ
Keel punctae 10-11 in 10 μ

Perceived in water bodies of paddy field as well as in cultures.

50. Nitzschia hantzschiana Rabenh.

Plate-XXXIX, Fig.50

Dimensions: Length - 37.8-42μ
Breadth - 2.8-4.8μ

Striae - 27-28 in 10μ
Keel punctae - 9-10 in 10μ

Perceived in water bodies of paddy field as well as in cultures.

51. Nitzschia intermedia Hantzsch

Plate-XXXIX, Fig.51

Dimensions: Length - 74.8-90μ
Breadth - 4.2-6.6μ
Striae - 23-24 in 10μ
Keel punctae 8-9 in 10μ

Perceived in water bodies of paddy field as well as in cultures.

52. Nitzschia kuetzingiana Hilse

Plate-XXXIX, Fig.52

Dimensions: Length - 11.2-15.0μ
Breadth - 4.2-5.6μ
Keel punctae - 13-14 in 10μ

Perceived in water bodies of paddy field as well as in cultures.

53. Nitzschia palea (Kg.) W. Sm.

Plate-XXXIX, Fig.53

Dimensions: Length - 22-58.9 μ
Breadth - 2.8-4.3 μ
Striae - 30-36 in 10 μ
Keel punctae 11-12 in 10μ

Perceived in water bodies of paddy field as well as in cultures.

54. Nitzschia parvula Lewis var. **terricola** Lund

Plate-XXXIX, Fig.54

Dimensions : Length -42-54 μ
Breadth - 5.6-7.0 μ
Striae - 30-32 in 10 μ
Keel punctae 6-7 in 10 μ

Perceived in water bodies of paddy field as well as in cultures.

55. Nitzschia vermicularis (Kg.) Hantzsch.

Plate-XXXIX, Fig.55

Dimensions: Length - 159-195.7µ
Breadth - 9.4 - 10.8µ
Striae - 32-34 in 10µ
Keel punctae 6-9 in 10µ

Perceived in water bodies of paddy field as well as in cultures.

Family - Surirellaceae
Sub-family - Surirelloideae
Genus - **Surirella** Turp.

56. Surirella baltica Schum.

Plate-XXXIX, Fig.56

Dimensions: Length - 36.5-37.9 µ
Breadth - 16.8-18.9 µ
Striae - 12 in 10 µ
Costae dots 2.5 in 10 µ

Perceived in water bodies of paddy field as well as in cultures.

Chapter 4

SUMMARY AND CONCLUSION

Paddy forms the major staple food of the Indian people. In Vidarbha, Soil-climate complex is favourable for paddy cultivation. Paddy field algae of Vidarbha Region have not beer thoroughly investigated hence the present attempt has been made.

Paddy field snows a variable environment for the growth of different types of algae at different seasons. Hence collection of algae and cultures of algae at different seasons were made to finalize the list of algae present in the paddy fields of Vidarbha Region of Maharashtra.

In all, 51 places in 8 districts of Vidarbha were taken for study of which 9 each from Nagpur, Bhandara and Chandrapur districts, 5 each from Wardha and Yavatmal, 4 each from Amravati and Buldhana districts and 6 places from Akola districts. Nine different culture medias were used to avoid the possibility of elimination of any alga from the list due to culture condition.

In all 500 algal taxa from paddy fields of Vidarbha were isolated and identified. Of these 323 were Cyanophyta,

90 Chlorophyta, 6 xanthophyta, 25 Euglenophyta and 56 Bacillariophyta.

From the paddy fields of Nagpur districts 211 taxa of algae could be identified. Out of these 152 belonged to Cyanophyta, 35 to Chlorophyta, 3 to Xanthophyta, 5 to Euglenophyta and 16 to Bacillariophyta. The pH of .soil ranged fron 6.8 to 8.0.

In all 132 a1ae taxa were identified from the paddy fields of Bhandara district from 9 different places. Out of these 62 belonged to Cyanophyta, 34 to Chlorophyta, 10 to Euglenophyta and 26 to Bacillariophyta. The report of the soil analysis showed a pH range from 6.6 to 7.6.

Nine different places of Chandrapur district were stuoie6 to know the paddy field algal flora of district. In all 137 algal taxa could be identified from these places. Out of which 87 belonged to Cyanophyta, 25 to Chlorophyta, 1 to Xanthophyta, 2 to Euglenophyta and 22 to Bacillariophyta. The pH range was observed to be 7.3 to 8.2.

Due to the rare occurrence of paddy cultivation in Wardha district only 5 different places could be selected for studies. The total al gal taxa identified numbered 182, which is higher than that of Bhandara and Chdrapur district. The comprise of 135 taxa of Cyanophyta, 21 taxa of Chlorophyta, 1 taxa of Xanthophyta, 7 taxa of Euglenophyta and 18 taxa of Bacillariophyta. The pH ranged from 7.4 to 8.1 which is more alkaline as compared to the places of other districts of Vidharbha.

The main crops of Amravati district too were other than paddy. Only very rarely the paddy cultivating fields could be traccd. Thus only 4 places could be taken for the studies. In all 185 algal taxa could be identified from these fields out of which 134 belonged to Cyanophyta, 26 to Chlorophyta, 1 to

Xanthophyta, 6 to Euglenophyta and 18 to Bacillariophyta. The soil pH of the district ranged from 7.9 to 8.1.

Total 5 places could be selected from Yeotmal district for investigation. The total algal taxa from the districts came out to be 167 to which 128 belonged to Cyanophyta, 22 to Chlorophyta, 3 to Euglenophyta and 14 to Bacillariophyta. The soil pH ranged from 7.0 to 7.7.

From Buldhana district 4 places were selected for the study. Total 1[illegible]3 algal taxa could be identified which are distributed amongst Cyanophyta (122), Chlorophyta (38), Euglenophyta (13) and Bacillariophyta (20). The pH range was found to be 7.6 to 8.2.

From Akola district 155 algal taxa could be identified from the studies of 6 different places. These taxa included Cyanophyta (103), Chlorophyta (20), Xanthopphyta, (3) Euglenophyta (5) and Bacillariophyta (24). The soil pH ranged from 7.2 to 8.0.

Soil samples were analysed for physical and Chemical Characters to find out correlation between the results of the analysis with the algal flora of the respective fields. In general it is observed that in Vidarbha region, paddy field exhibit alkaline condition of soil. Naturally cyanophyta predominates over other groups of algae. Presence of some fresh water chlorophycean and bacillariophcean members in the paddy fields may be due to water condition of the soil in rainy season.

The algal flora of a particular region or crop field depends on the climate, of the region, environment of the field and nature of cultivation. The interaction between the algal flora and the crop plant in a crop field. Similarly paddy too have much important effects by the algal flora of the paddy field.

Paddy field shows a variable environment for the growth of different types of algae at different seasons. Hence collection of algae and cultures of algae at different seasons were made to finalise the list of algae present in the paddy fields of Vidarbha region of Maharashtra.

In all 51 places in 8 districts of Vidarbha were taken for study of which 9 each from Nagpur, Bhandara and Chandrapur districts, 5 each from Wardha and Yavatrnal, 4 each from Amravati and Buldana districts and 6 places from Akola district. Nine different culture medias were used to avoid elimination of an alga from the list due to culture condition.

Soil analysis of the respective fields were made to correlate the presence of an alga in a particular type of soil.

In all 500 taxa from paddy fields of Vidarbha were isolated and identified. Of these 323 were Cyanophyta, 90 were Chlorophyta, 6 were Xanthophyta, 25 were Euglenophyta and 56 were Bacillariophyta.

From the paddy fields of Nagpur districts 211 taxa of algae could be identified. Out of these 152 belonged to Cyanophyta, 35 belonged to Chlorophyta, 3 belonged to Xanthophyta, 5 belonged to Euglenophyta and 16 belonged to Bacillariophyta. The pH of soil ranged from 6.8 to 8.0. Thus the soil of paddy fields of Nagpur district being alkaline the dominance of the Cyanophycean forms over other groups of algae was not unexpected.

In all 132 taxa were identified from the paddy fields of Bhandara district. Total 9 different places of the district were studied. These algal taxa belonged 62 to Cyanophyta, 34 to Chlorophyta, 10 to Euglenophyta and 6 to Bacillariophyta. The report of the soil analysis showed a pH range from 6.6 to 7.6 to state an alkaline condition.

Of course, due to the alkaline environments, the Cyanophyta dominated, but the presence of some of the fresh water Chlorophycean members and Bacillariophycean members are due to water logged condition of the field in rainy season.

Nine different places of Chandrapur district were studied to list the paddy field algal flora of the district. In all 137 algal taxa could be identified from these places, out of which 87 belonged to Cyanophyta, 25 belonged to Chlorophyta, one belonged to Xanthophyta, 2 belonged to Euglenophyta and 22 belonged to Bacillariophyta. The pH range was observed to be 7.3 to B.2, hence alkaline environment which explains the dominance of the Cyanophyta over groups of algae.

Due to the rate occurrence of paddy cultivation in Wardha district only 5 different places could be selected for studies. Still the total algal taxa could be identified reached upto 182, higher than that of Bhandara and as well as Chandrapur district. In these 182, 135 taxa of Cyanophyta, 21 taxa of Chlorophyta, 1 taxa of Xanthophyta, 7 taxa of Euglenophyta and 18 taxa of Bacillariophyta were present. The pH ranged from 7.4 to 8.1 which is more alkaline as compared to the places of other districts. Hence naturally, the Cyanophyta exceeds the highly over other groups of algae.

The main crops of Amravati district too were other than paddy. Only very rarely the paddy cultivating fields could be traced. Thus only 4 places could be taken for the studies. In all 185 algal taxa could be identified from these fields out of which 134 belonged to Cyanophyta, 26 belonged to Chlorophyta, 1 belonged to Xanthophyta, 6 belonged to Euglenophyta and 18 belonged to Bacillariophyta. The pH of the district ranged from 7.9 to 8.1 which clearly indicates the dominance of Cyanophyta over other groups of algae.

Total 5 places could be selected from Yeotmal district for investigation. The total algal taxa from the districts came out to be 167 of which 128 belonged to Cyanophyta, 22 belonged to Chlorophyta, 3 belonged to Euglenophyta and 14 belonged to

Bacillariphyta. The pH ranged were found to be 7.0 to 7.7. The alkaline condition indicates the dominance of Cyanophyta.

From Buldana district 4 places were selected for the study. Total 193 algal taxa could be identified which are distributed amongst Cyanophyta (122). Chlorophyta (38), Euglenophyta (13) and Bacillariophyta (20). The pH range was found to be 7.6 to 8.2 which explains the dominance of Cyanophyta over other groups of algae.

From Akola district 155 algal taxa could be identified from the studies of 6 different places. These taxa included Cyanophyta (103), Chlorophyta (20), Xanthophyta (3), Euglenophyta (5) and Bacillariophyta (24). The pH ranged between 7.2 to 8.0, which explains the dominance of Cyanophycean members over other groups.

REFERENCES

ALEKSANDROVA, I.V., 1951. The process of humus formation in several primitive mountain soils. Pochvovendnia, 1951 : 604.

ALEXANDER, V., 1975. Nitrogen fixation by blue green algae in Polar and subpolar regions in *'Nitrogen fixation by free living micro-organisms'* (W.D.P. Stewart Ed.). Cambridge University Press, Cambridge, pp.175-l88.

ALl, S. AND SANDHU, G.R., 1976. Laboratory cultivation of blue-green algae of saline soils and rice fields. In *'proceedings of the National Panel on Algae and its utilization.'* Institute of Public Health Engineering and Research, University of Engineering and Technology, Lahore, Pakistan.

ALLISON, F.E., HOOVER, S.R. AND MORRIS, B.J., 1937. Physiological studies with nitrogen fixing alga Nostoc muscorum. Bot. Gaz., **98** : 433.

AMMA, P .A., AIYER, R. S. AND SUBRAMONEY, N., 1967. Occurrence of blue-green algae in the arid soils of Kera1a. Agrl. Res. Jour. Kera1a, 4 (2) : 140-143.

ANANTANI, Y. S. AND MARATHE, K. V., 1972a. Observations on the algae of same arid and semi-arid soils of Rajasthan. J. Univ. Bombay XLI (68) : 88-93.

ANANTANI, Y.S. AND MARATHE, K.V., 1972 b. Soil aggregating effects of some algae occurring in the Soils of Kutch and Rajasthan. J. Univ. Bombay XLI (68): 94-100.

BANARJI, J.C., 1935. On algae found in soil samples from an alluvial paddy field at Faridpur. Bengal Sci. Cult., 1 : 298-299.

***BEIJERINCK, M.W., 1893.** Bericht uber meine kulturen miederer Algen auf Nahregelatine. zbl. Bakt., 13 : 368.

BHARATI, S.G. AND BHONGALE, U.D., 1975. Soils and soil algae of Karnataka State. I. Soils and Soil Algae of Dharwar. J. Karnataka Univ. Sci., 20B : 142-150.

BONGALE, U.D., 1979. Cyanophyceae from the cultivated soils of Karnataka State, India. Proc. 66th Ind. Sci. Cong. Abst., 1 : 1.

BOOTH, W.E., 1941. Algae as pioneers in plants succession and their importance in erosion control. Ecology, 2: 38-46.

BRISTOL, B.M., 1919. On the retention of vitality by algae from old stored soils. New Phytologist, 18 : 92-107.

BRISTOL, B.M., 1920. On the algal flora of some desiccated English soils. An important factor in soil biology. Ann. Bot., 34 : 35-80.

BRISTOL, B.M., 1926. On the relation of certain soil algae to some soluble carbon compounds. Ann. Bot. (London), 40 :147-201.

BRISTOL, H.M., 1927 a. On the algae of some normal English Soils. J. Agril. Sci. Camb., 17 : 563-588.

BRISTOL, B.M., 1927 b. On the carbon nutrition of sane algae isolated from soil. Ann. Bot. (London), **41** : 509-517.

BRISTOL, B.M., 1928. One of the influence of light and glucose on the growth of soil algae. Ann. Bot. (London), **42** : 317-345.

BRISTOL ROACH, B.M., 1927. The present position of our knowledge of the distribution and functions of algae in the soil. I[st] Inter. Cong. Soil Sci. Washington, D.C., 30-38.

BROCK, T.D., 1973. Lower pH limit for the existence of blue green algae evolutionary and ecological implications. Science (Wash), **179** (4072) 480-483.

BUNT, J.S., 1961. Nitrogen-fixing blue-green algae in Australian rice soils. Nature, Lond., **192** : 480-490.

CHODAT, F., 1928. Sur. La. specifitive des stichococcus du sol du pare National. Comp. rend Soc. Phys. Nat. Geneve. **45**: 26.

CHAUDHARI, P.R., 1977. Studies on the microflora (soil) of some orange archords around Nagpur. Ph.D. Thesis, Nagpur University, Nagpur.

DAS, V., MITRA, A.K. AND LALORAYA, V.K., 1974. Studies on soil algae from certain hills of India. Beinheft, 47. Zur. Nova. Hedwigia : 103-132.

DE, P.K., 1939. The role of blue-green algae in nitrogen fixation in rice-fields. Proc. Roy. Soc., Lond. **(13)** : 127-139.

DE HALPERIN DELTA, R., 1976. Cyanophyceae of algal crusts from chaco and fannosa provinces (Argentina) Physics SE cc. c. cont. ORG. TERR., 35(90) : 1-16 (In Spanish).

DE HALPERIN DELTA, R., MARIA, C.Z., DE MULE, GLORIA, Z. AND De CAIRE, 1976. Algal crusts as sources of nitrogen in subhumid and semi-arid soils (chaco and farmosa provinces, Argentina (in Spanish) Darwiniana (B.Aires) **20(3/4)** : 341-370.

DHARMAJIRAO, 1968. Nitrogen fixing blue green algae in Berhampur (Orissa) Soils. Proc. 6th Ind. Sci. Cong. No.7: p.3.

DOOLEY, F. AND HOUGHTN, J. A., 1973 a. The nitrogen fixing capacities and the occurrence of blue-green algae in peat soils. Br. Phycol. J., **8**(3) : 283-289.

DOOLEY, F.. AND HOUGHTN, J .A., 1973b. The physiology of some blue-green algal isolates from peat soils. Br. Phycol. J.S., (3) : 295-300.

DURRELL, L.W., 1959. Algae of Colorado Soils. Amer. Mid. Nat., **21** : 322-328.

DURRELL, L.W. AND SHEILDS, L.M., 1962. Algae of Death Valley. Trans. Amer. Micros. Soc., **81** : 267-273.

DUTTA, N., VENKATRAMAN, G. S., 1958. An exploratory study of the algae of some cultivated and uncultivated soils. Indian J. Agron., **3** : 109-115.

EL-NAWAWY, A. S. AND HAMDI, Y. A., 1975. Research on blue-green algae in Egypt, 1958-1972. In *Nitrogen fixation by free living micro-organisms* (Ed. W. D.P. Stewart). Cambridge Univ. Press, Cambridge.

ENGLUND, B., 1976. Nitrogen fixation by free-living microorganisms on the lava field of Heimaey Iceland. Oikos 27(3): 428-432.

FLETCHER, J. E. AND MARTIN, W. P . 1984 Some effects of algae and molds in the rain-crust of desert soils. Ecology, **29**: 95-100.

FOGG, G.E., 1951. Studies on nitrogen fixation by blue green algae II. Nitrogen fixation by Mastigoladus imnosus Cohn. J. Expt. Bot., **2** : 117.

FOGG, G.E., 1960. Recent advances of our knowledge of the nitrogen fixation of blue-green algae. Proc. Symp. Algology. I.C.A.R., p.115.

FOGG, G.E., 1962. Extracellular products. "Physiology and Biochemistry of algae" Academic Press, New York and London, 475-486.

FOGG, G.E. AND STEWART, W.D.P., 1965. Nitrogen fixation in blue-green algae. Sci. Progr., **53** : 191-201.

FRITSCH, F.E., 1907 a. The role of algal growth in colonization of new ground and in the determination of scenary. Geographical Jour., **30** : 531-548.

FRITSCH, F.E., 1907 b. The subarial and freshwater algal flora of the Tropics. Ann. Bot. Lond., **10** : 235-275.

FRITSCH, F.E., 1922 a. The terresterial alga. Jour. Eco., **10**: 220-236.

FRITSCH, F.E., 1936. The role of the terresterial algae in nature. Essays in Geobotany in honour of W.A. Setchell 195-217. Univ. Calif. Press.

FRITSCH, F.E. AND DE, P.K., 1938. Nitrogen fixation by blue-green algae. Nature, **142** : 878.

FRITSCH, F.E. AND HANES, F.M., 1923. The moisture-relations of terrestrial algae II. The changes during exposure to drought and treatment with hypertonic solutions. Annals. Bot., **37** : 683-728.

FRITSCH, F.E. AND JOHN, R.P., 1942. An ecological and taxonomic study of the algae of British Soils. II consideration of the species. Ann. Bot. N.S., **6**: 373-395.

FULLER, W.B. AND ROGERS, R.N., 1952. Utilization of the phoshorus of the algal cells as measured by the Neubauer technique. Soil. Sci., 74 : 417-430.

GISTIL, R., 1932. Zur kenntins der. Erdalgen. Arch. Microbiol. **3**: 634-649.

GONZALUES, E.A. AND GANGLA, K. S., 1949. Observations on the algae of paddy field soils. J. Univ. Bombay, **18(3)** : 51-59.

GONALVES, B.A. AND YALAVIGI, V .S., 1960. Algae in the rhizosphere of some crop plants. Proc. Sym.. Algology, New Delhi pp.335-344.

GRAEBNER, P. 1895. Studi enuber die norddeutsche Heide Engl. Bot. Jahrb., **20** : 500 (quoted by Petersen, 1935).

GRANHALL, U., 1970. Acetylene reduction by blue-green algae isolated from Swedish Soils. Oikos, **21** : 330-332.

GRANHALL, U. AND HENRIKSSON, E.,1969. Nitrogen fixing blue-green algae in Swedish soils. Oikos, **20**: 175-178.

GUPTA, A.B., 1966. Algal flora and its importance in the economy of rice fields. Hydrobiologia, **28** : 213.

GUPTA, R.S. AND KUMAR, H.D., 1968. Blue-green algal flora of Udaipur and its neighbourhood. Rev. Algol., **9(2)** : 91-93.

HARRISON, W.H. AND AIYER, P.A.S., 1913. The gases of swamp rice soils, their composition and relation to the crop Mem. Dep. Agric. India (Chem.Ser.) **3** : 65.

***HUBER - PESTALOZZI, G., 1955.** Die Binnengewasser Bd., 16. Teil 4. Das phytoplankton des Susswassers. Euglenophycean. Stuttgart.

JAMES, E.J. 1935. An investigation of the algal growth in some naturally occurring soils. Beiheste zum Botanischen. Zentralblatt., 53 : 519-553.

JOHN, R.P., 1942. An ecological and taxanomical study of algae of British soils. I. The distribution of the surface growing algae. Ann. Bot. N.B., **6**: 323-349.

JONES, K., 1977 a. Acetylene reduction by blue-green algae in subtropical grassland. New Phytol., 78 (2) : 421-426.

JONES,K., 1977 b. The effect of light intensity on acetylene reduction by blue-green algal mats in sub-tropical grasslands. New Phytol., 78(1) : 427-431.

JONES, K., 1977 c. The effect of temperature on acetylene reduction by mats of blue-green algae in subtrcpical grass-land. New Phytol., 78(2) : 435-436.

JONES, K., 1977 d. Acetylene reduction by mats of blue-green algae in sub-tropical grass lands possible contribution by other micro-organisms. New Phytol., 79(2) : 437-440.

JURGENSEN, M.F. AND DAVEY, C.B., 1968. Nitrogen fixing blue-green algae in acid, forest and nursery soils. Can. J. Microbiol., 14 : 1179-1183.

KAMATH, N.D. AND PATEL, M.Z., 1973. Soil algae of a rice field at different depths. Botanique, Vol.IV No.2 pl.

KASZUBAIK, H., 1976. The influence of mineral fertilizers on micro-organisms in various conditions of the soil environment II The influence of mineral fertilizers on some groups of micro-organisms at different soil moisture. Pol. J. Soil Sci., 9(1) : 41-46.

KHAN, S.I., 1957. Biologia (Pakistan) 3 : 29.

KHAN, T.N. AND MATHUR, A., 1976. Algal flora of the rice fields around Dehradun, India. Rev. Algol. 1(3/4) : 333-338.

KHOJA, T. AND WHITTON, B.A., 1971. Heterotropic growth of blue-green algae. Arch. Mikrobiol., 79(3) : 280-282.

KHURANA, A. S. AND VENKATARAMAN, G. S., 1968. Algal and fungal flora of paddy soils. Indian J. Microbial., 8(2): 91-94.

KING, J.M. AND WARD, C.H., 1977. Distribution of edaphic algae as related to land usage. Phycologia, **16(1)** : 23-30.

LALORAYA, V.K. AND MITRA, A.K. 1974 a. Studies on the blue-green algae of the paddy fields of India. I. Cultural studies general considerations and distribution pattern of the blue-green algae in the paddy fields of India. Nova. Hedwigia.. Beihert, **47:** 263-338 (ed. J.Crame).

LALORAYA, V.K. AND MITRA, A.K. 1974 b. Studies on the blue-green algae of the paddy fields of India, Part II Taxonomic considerations of the blue-green algae obtained from paddy fields of India. Nova Hedwigia, Beiheft, 47 : 263-338 (ed.J.Crame).

LALORAYA, V.K. AND MITRA, A.K. 1974 c. Studies on the blue-green algae of the paddy fields of India. Part IV. New members of Nostocaceae and Rivulariaceae. Nova Hedwigia Beiheft 47 : 347-359 (ed.J.Crame).

LEHRI, L.K. AND MEHROTRA, C.L., 1970. Note on the effect of artificial inoculation of Aulosira fertilissima in rice crop. Indian J. Agri. Sci., **40(1)** : 65-67.

LUND, J.W.G.. 1947. Observations on soil algae. II. Notes on groups. other than diatoms. New Phytol., **46** : 35-60.

LUND, J.W.G., 1962. Soil algae in : *Physiology and Biochemistry of Algae* (Lewin, R.A. ed.) Academic Press, New Yark and London, pp.759-770.

MAHAPATRA, I.C., MUDHOLKAR, N.J. AND PATTNAIK, H., 1971. Effect of N, P and K. fertilizers on the prevalence of algae under field conditions. Indian Jour. Agronomy 16(1) : 19-22.

MARATHE, K.V., 1963. A study of the effect of fertilizers on the algal flora of paddy field soils from Karjat. J. Univ. Bombay, **31** (3 and 5) : 1-10.

MARATHE, K.V., 1964. A study of the effects of green manure on the Subterranean algal flora of paddy field soils. J. Biol. Sci., 7 : 1-7.

MARATHE, K.V., 1965. A study of the subterranean algal flora of some mangrove swamps. J. Indian Soc. Soil, Sci., **13(2)**: 81-84.

MARATHE, K.V., 1966. Taxonomic and ecological studies of subterranean algae from cultivated fields of Maharashtra. Ph.D. Thesis, University of Bombay.

MARATHE, K.V., 1967. Studies on soil algae cf India. 1. Soil algae from cultivated fields of Dhulia (Maharashtra State). J. Univ. Bombay, **35** (3 and 5) : 100-105.

MARATHE, K.V., 1969. Studies on soil algae of India II. Soil algae from the cultivated fields of Jalgaon. J. Univ. Bombay, **38.** : 69-72.

MARATHE, K.V., 1970. Role of soil algae in water conservation. J. Biol. Sci., **38** : 1-8.

MARATHE, K.V., 1972 a. Role of some blue-green algae in soil aggregation. In : *Taxonomy and Biology of blue-green algae* (ed.T.V. Desikachary) 328-331.

MARATHE, K.V., 1972 b. Soil algae of same Indian arid soils (Rajasthan). J. Univ. Bombay **XII (68)** : 88-93.

MARATHE, K.V., AND KHUSHALDAS,C., 1975. Observations on the algal crusts of Nagpur soils. J. Univ. Bombay, XLIV : 66-76.

MARATHE, K.V. AND NAVALKER, B. S., 1983. A study of the effect of fertilizers and manures on the subterranean algal flora of paddy fields. Soils. Jour. Bombay Univ., XXXI (3 and 5) : 11-19.

MARTIN, T.L., 1940. Proc. 3rd Int. Congr. Microbiol., N.Y. 697.

MARTIN, T.L., 1949. Occurrence of algae in some virgin utah soils. Soil Sci. Soc. Amer. Proc., 4 : 249-250.

MARTIN, J.P., MARTIN, W.P., PAGE, J.B., RANEY, W.B. AND DEMONT, J.D., 1955. Soil aggregation. Advan. Agronomy, 7: 1-37.

MATHUR, K.S. AND GOUR, A.C., 1977. Effect of humus and microbial inoculants on crop yield and nitrogen uptake. Microbiologia **46(1)** : 149-154.

MITRA, A.K., 1951. The algal flora of certain Indian soils. Indian J. Agri. Sci. **21** : 357.

NAEGELI, C., 1849. Gattungen einzelliger Algen, physiologisch and systematisch bearbeitet. Naye Denkschriften Allg. Schweiz Natur. Ges., 12 : 139, pp. Zurich.

PANDEY, D.C., 1965. A study of the algae from paddy field soils of Ballia and Ghazipur districts of Uttar Pradesh, India 1. Cultural and ecological considerations. Nova Hedwigia 9 : 297.

PETERSEN, J.B., 1934. The algal vegetation of Hammer Bakker. Botanisk Tiddskrift, 42 : 1-48.

PETERSEN, J.B., 1935. Studies on the biology and taxonomy of soil algae. Dansk bot. Ark **8(9)** : 1-80.

PIPER, C.S., 1950. Soil and plant analysis. Univ. of Adelaide, Hans Publishers, Bombay.

PRASAD, B.N. AND MEHROTRA, R.K., 1978. Some new additions to cyanophycean flora of India. J. Indian Bot. Soc., 57 : 98-101.

PRASAD, B.N., MEHROTRA, R.K. AND YASHPALSINGH, 1978. On pH tolerance of some soil blue green algae Acta. Botanica Indica, **6** : 130-138.

PRASAD, B.N. AND SRIVASTAVA, P.N., 1968. Systematic and ecological studies on algae of alkaline usar soils. Phykos.,7 (**1 and** 2): 102-111.

RAJARAM MOHAN RAO, 1976. Nitrogen fixation as influenced by moisture content, ammonium sulphate and organic sources in paddy soil. Soil Biol. Biochem. **8**(5) : 445-448.

REDDY, K.V.S., 1979. The Blue-green algae of some cultivated fields of Vidarbha.. Ph.D. Thesis, Nagpur University, Nagpur.

RENAUT, J., SASSON, A., PEARSON, H.W. AND STEWART, W.D.P., 1975. *Fixation by Free living Micro-organisms* (ed. W.D.P., Stewart) Cambridge Univ. Press, Cambridge, pp.229-248.

ROGER, P. AND JACQ, 1972. Rapport preli.min aire concernant unantagonisme entredes Bacteries sulfato reductrices et une souche de cyanophycee dans la rhizispere etla spermosphere du riz. Orstom. Daker. Pub roneo diff. Limitee.

ROGER,P. AND REYNAUD, P., 1976. Algal population dynamics during a growing season in a senegles rice field. Rev. Ecol. Biol. Soc., **13**(4) : 545-560.

SANKARAM, A., 1971. Work done on blue green algae in relation to agriculture. I.C.A.R., New Delhi: **28**.

SHIELDS, L. M. AND DURELL, L. W., 1964. Algae in relation to soil fertility. Bot. Rc., **30** : 92-128.

SHTINA, E.A., 1957. The development of soil algae in Sod. podzolic soils. Pochvovedenie : 12-18.

SHTINA, E.A., 1969. The results and tasks of experimental studies in algology. Tr. Kirov. Seliskokhoz. Inst., **20**(**40**) : 3-22.

SHTINA, E.A., 1972. Some pecularities of the distribution of nitrogen fixing blue-green algae in soils. In "Taxonomy and Biologw of Blue-green algae". (Ed. T.V. Desikachary) University of Madras - India, pp.294-295.

SINGH, H.D., 1933. A contribution to our knowledge of the algal flora of Lahore soils. J. Indian Bot. Soc., **12**: 102.

SINGH, R.N., 1939. An investigation .in to the flora of paddy field soils of the united provinces. 1. Indian J. Agri. Sci., **9(1)**: 55.

SINGH, R.N., 1942. The fixation of elementary nitrogen by some of the commonest blue-green algae from the paddy field soils of the united provinces. Indian J. Agri. Sci., **12** : 743-756.

SINGH, R.N., 1950. Reclamation of usar lands in India through blue-green algae. Nature, **165**: 325-326.

SINGH, R.N., 1961 a. Reclamation of usar lands in India through blue green algae. Nature, **165** : 325-326.

SINGH, R.N., 1961 b. Role of blue-green algae in nitrogen economy of Indian Agriculture. I.C.A.R., New Delhi.

SINHA, J.P. AND PANDEY, S., 1972. Blue green algae of paddy fields chota Nagpur II. Proc. 59th Ind. Sci. Cong. No.**14** p.295.

SINGH, P.K., 1974. Effects of pesticides on blue-green algae. Arch. Microbial., **89** : 317-320.

STEWART, W.D.P., 1963. Nitrogen fixation and production of extra cellular nitrogeneous substances by an endophytic Nostoc strain isolated from the root nodules of Egyptian clover (Trifolium alexandrium). Proc. Symp. Algologw, I.C.A.R., p.119.

STEWART, W.D.P., 1970. Algal fixation of atmospheric nitrogen. Pl. Soil, **32** : 555-588.

SUBBA RAJU, 1967. An ecological and taxonomic study of the algal flora of certain soils from Andhra Pradesh, India. Ph.D. Thesis, submitted to the Osmania University, Hyderabad.

SUBBA RAJU, 1972. The blue-green algae from the soils of India. In "*Taxonomy and Biology of Blue-green Algae*" (ed. T.V. Desikachary): 332-352.

SUBRAHMANYAM, A., 1972. Algal flora of Bastar India : blue-green algae of Jagdalpur. Rev. Algol., **10**(4) : 290-296.

SUBRAHMANYAM, R., RELWANI, L.L. AND MANNA, G. B., 1965. Fertility build up of rice field soils by blue green algae. Proc. Indian Acad. Sci., **62 B** : 252-272.

TELL, G., 1976. Soil algae from around Euenos Aires (Argentina). Darwiniana (B.Aires) **20**(3/4) : 491-548.

TIWARI, G.L., 1972. A study of the blue-green algae from paddy field soils of India. Hydrobiologia, **39**(3) : 335-350.

TRAINOR, F.R., 1970. Survival of algae in a desiccated soil. Phycologia, 9(2) : 111-113.

VENKATARAMAN ,G.B., 1961. Studies on nitrogen fixation by *Cylindrospermum sphaerica* Prasad under various conditions. Proc. Natn. Acad. Sic. India, **31** : 100-104.

VENKATARAMAN, G.S., 1972. *Algal Biofertilizers and Rice cultivation.* Today and tomorrows Printers and Publishers, New Delhi, p.75.

VENKATARAMAN , G. S., 1975. The role of blue-green algae in tropical rice cultivation. In *"Nitrogen fixation by free living micro-organisms".* (ed. W.D.P. Stewart) Cambridge University Press, Cambridge, pp.207-2l8.

VENKATARAMAN, G.S., JACOB, K.M. AND GOYAL, S.K., 1964. Nitrogen fixation by the endophytic alga from the corolloid roots of cycas revoluts. Proc. Nat. Acad. Sci., India, **34** : 153.

VENKATARAMAN, G.S. AND NEELKANTAN, S., 1967. Effect of cellular constituents of the nitrogen-fixing blue green alga *Cylindrospermum muscicola* on the root growth of rice plants. J. Gen. Appl. Microbiol., **13** : 53.

VENKATARAMAN, G.S. AND SAXENA, H.K., 1963. Studies on nitrogen fixation by blue-green algae 1 V. Liberation of free aminoacids in the medium. Indian J. Agric., Sci., **33(1)**: 77.

WAKSMAN, S.A., 1952. Soil Microbiology. John Wiley and Sons Inc., New York, p.356.

WATANOBE, A., 1951. Production in culture solution of some amino acids by the atmospheric nitrogen fixing blue-green algae. Arch. Biochem. Biophys., **34** : 50.

WATANABE, A., 1961. List of algal strains in collection at the Institute applied microbiology Univ. of Tokyo. J. Gen. Microbial., 6 : 283.

WATANABE, A. I 1965. Studies on the blue-green algae as green manure in Japan. Proc. Nat. Ac. Sci. India, **35**.

WATANABE, A., 1970. Studies on application of cyanophyta in Japan. Schweiz, Z. Hydrol., **32** : 566-567.